Beaches,
Bureaucrats,
and
Big Oil

Beaches,
Bureaucrats,
and Big Oil

Garry Mauro

Library of Congress Cataloging-in-Publication Data is available.

ISBN 0-96583441-7 *(paper)*
ISBN 0-96583344-0-9 *(cloth)*

Published by
Look Away Books
101 West Sixth Street
Suite 609
Austin, Texas 78701

Cover photograph by Scot Hill
Cover design by Martin Erhart
Book design by Barbara Whitehead

Printed in Texas

Printed on Recycled Paper

This book is dedicated to 186,000 Adopt-A-Beach volunteers, who have given their time and energy for the past 11 years to keep Texas beaches clean;

to the parents of Texas school children, who give countless days, nights, and weekends to make Texas schools a better place for our children; and

to the generations of Texans who built this state by using its natural resources to produce a special quality of life and who left this state and its resources better than they found them.

Acknowledgments

This book is based on core values that have guided me since my days in student government at Texas A&M. It reflects a vision for Texas that I've sharpened during fifteen years as land commissioner.

Beaches, Bureaucrats, and Big Oil is the product of more than a year's work with writers and friends. Professionals helped shape my words into this book—but these are my words.

Jan Reid wrote speeches for me during the 1980s, has written three books, and is a former senior writer for *Texas Monthly*. When I needed an editor to make this dream a reality, Jan was the obvious choice.

Bob Mann is spokesman for the oil spill division of the Texas General Land Office. He has been a press secretary to several national figures and taught journalism at Southern Methodist University. A longtime friend, Bob took leave without pay from the Land Office to draft the original manuscript.

Joe Cutbirth, a former political reporter for the *Fort Worth Star-Telegram,* manages marketing projects at the Land Office. Joe Holley is an author and former *Texas Observer* editor who writes speeches for me. In their spare time they helped shape this book.

At the Land Office, I work with some of the brightest and most talented people in state government. They are responsible for every success I've had since I took office. I can't ever thank them all by name. I thought this was a book that deserved to be written. But the story belongs to them.

Contents

PART I

CLEAN BEACHES

1

Dirty Secrets

I never would have dreamed that a decision to go to a 1986 beach cleanup and risk missing a Texas A&M–Baylor football game would ultimately make me the enemy of international oil companies, the plastics and shipping industries, the United States Navy brass, and lobbyists and bureaucrats throughout the land. I thought it was just a photo opportunity.

I was going to scoot down to Port Aransas that Saturday morning, pick up some beer cans and other litter, try to generate some positive press—and then get the heck back to what many consider the best college football game of the 1980s. Go figure.

By picking up a little trash on an isolated part of Mustang Island, I was starting on a journey that would take me down the blind alleys of the Austin and

Washington bureaucracies, into the secret struggles of the most potent lobby web in America, across the ocean to the tense formality of international politics in London, stopping along the way to contend with the pariahs and paranoia of Cuba.

With some creative scheduling and a pilot willing to fly early and fast, I agreed to a trip to the Texas coast that would have a profound impact on me—and shape my outlook and public policy decisions for the next decade and beyond.

You see, I'm a Texas Aggie, through and through. I'm descended from Italian immigrants who, early in this century, were enticed to settle in the Brazos River valley. Two of my great-grandfathers arrived in this country at Ellis Island. One made his way to New Orleans, then Houston, and then to Bryan, Texas; the other landed at Galveston, Texas, and made the trip inland. Promoters sold farmland in the river bottom to some of the Sicilian immigrants at ten cents an acre. The Brazos flooded for weeks at a time in those years. They had to build their houses on stilts, and the air sang with mosquitoes. It wasn't a great place to live. But later the Brazos was dammed far upstream, and the silted bottomland became some of the richest farming country Texans had ever seen. The Italians weren't so dumb after all.

My Mauro great-grandparents were tenant farmers who eventually managed to buy 180 acres. When my great-grandfather decided to move into Bryan, my great-grandmother didn't want to give up her house, so they just took it with them; neighbors pitched in and helped them move it. His son, my grandfather, opened a grocery store across the street from my great-grandfather Cash. My great-grandfathers lived within three blocks of each other. I've been told that in 1910 Brazos County had the second highest concentration of Italian immigrants in the country, and my memories of how they lived, even forty years later, are like a scene in some village. The old house sat on three lots. There were fig trees and a vegetable garden that would have made our ancestors proud.

My dad, Frank, was the first in the Mauro family to graduate from college—Texas A&M, of course. He got his degree in petroleum engineering. Dad was also the first of the Mauros to marry someone outside the Italian community. To some of his neighbors and kin, that was almost unthinkable. My mother, Lou, grew up in Alabama before moving to Dallas. Her dad had a third-grade education; he started out as a cucumber buyer and worked his way up to become a vice president for the WW Pickle company.

You're never going to hear me engage in immigrant-bashing. I was born in Bryan in 1948 and have three younger sisters, Sandy, Patti, and Mary Lou. For us it was an upbringing steeped in the value of hard work, the incredible opportunity in this country, the freedom to strive and dream. When I was a baby my dad worked in Oklahoma for Continental Oil. My grandparents urged him to come home and help run the family business, and he did for a while. Then he became a civil servant, managing post exchanges of Air

Force bases in Bryan, then Midwest City, Oklahoma, then Abilene, where he got back into the oil business, this time working in the marketing department of Gulf. That took us to Fort Worth, to Hurst, then to Waco. But during school holidays and summer vacations, I always wanted to be with family in Bryan. Naturally, I grew up captivated by all the ruckus just down the road in College Station.

In Waco I went to Reicher Catholic High School and played football in the mid-sixties. I was a guard. Later, Coach Gene Stallings made my youthful dream come true; he gave me a chance to play at Texas A&M for his Texas Aggies. I didn't have a scholarship, but the coaches helped me get a part-time job, and I lived in the athletic dorm. I got hurt, which was probably inevitable. I'm five-ten and weigh 180 pounds. I told Coach Stallings I wanted to bear down on my studies—an explanation he liked a lot better than that I was just quitting. Over the years, as he went on to the Dallas Cowboys, the Arizona Cardinals, the University of Alabama, and now home to Paris, Texas, I've stayed in touch with Gene Stallings. He's still one of my heroes.

I worked at the Pizza Hut and became an Aggie yell leader. Like all colleges in those years, A&M was feverish with politics, and the number of future elected leaders on campus was remarkable—Henry Cisneros, John Sharp, Chet Edwards, Rick Perry, Kent Caperton. I ran for student body president one year against the Corps of Cadets candidate Al Reinert, which is ironic given his bohemian style in later years. Al became a writer and moviemaker specializing in the NASA manned space program; he's been nominated for two Academy Awards and he co-wrote the screenplay of *Apollo 13*. Even as an Aggie politician, Al was a bit of a space cadet. He beat me narrowly, but it turned out he didn't have the necessary scholastic credits. I won't have the office this way, I announced: Put up another candidate. So they did—Gerry Geistweidt, who later served in the legislature as a Republican—and this time I got trounced. Live and learn. My college years were one of the best times in my life.

The Battle of the Brazos, as the A&M-Baylor game is called, is played every year in one of my hometowns. I wasn't about to miss the '86 game. No way. But there was just one conflict. Politics.

In a few weeks, I hoped to be elected to my second term as Texas Land Commissioner. In a nutshell, my job is to protect and manage the public lands in Texas and to administer the Veterans Land Board loan programs. But the novel scope and design of the office enable me to advance the priorities that underscore my philosophy of government: education, economic security, equal opportunity, and the environment. Notice I use the present tense.

I won that 1986 race with better than sixty percent of the vote, and two elections later, I am still the Texas Land Commissioner. Have been now for fifteen years. I'm aware that makes me the kind of politician a lot of people love to hate. But like I told delegates to the 1996 Democratic National Convention, which my old friend from Arkansas, President Bill Clinton, let me address for three minutes. I love my job.

Any officeholder who takes an election for granted is looking for a new line of work. So on that football Saturday more than ten years ago, I had to do some politics.

Mike Connolly, the public information director of the Texas General Land Office, had urged me to meet with Linda Maraniss, with whom he had lunched earlier in the month. Mike was a transplanted Washington, D.C., journalist and a good friend as well as a valued aide. He had covered the White House and once interviewed the genocidal Cambodian dictator Pol Pot, which must have been a strange experience. I listened to Mike, and he never hesitated to speak his mind, especially on how I ought to spend my time.

Linda, who is originally from Wisconsin, was then the director of the Gulf Coast Regional Office of the Center for Environmental Education. She had a proposal that Mike thought I should consider.

Although I had met Linda before, I did not know her well. I knew her husband a lot better. David Maraniss manned the Austin bureau of the *Washington Post*, and he was showing Texans the talent and instinct that would later win him the Pulitzer Prize when he reported on a brash Arkansas governor who thought he could actually beat George Bush.

Linda talks a lot faster than I do, and she got right to the point. "When we moved here," she told me, "I was really looking forward to it. Everybody said, 'Oh, you're going to love Texas. But the coast is a mess.' They were right." Linda wanted my support, and that of the General Land Office, for a beach cleanup that the Center for Environmental Education had started with a one-time grant. The theme of the cleanup was "Be a Beach Buddy."

Linda saw my role—and so did I, at the time—as something of a high-powered publicist. She had learned the hard way that her citizens' advocacy group could call a press conference for the most worthy cause, and still only a handful of reporters would show up. Even if she was married to one of the best of them. On the other hand, I could announce a press conference and, most of the time, it would draw a respectable crowd. And for good reason. Texas is hotly contested turf for journalists, and to ignore a press conference of a statewide elected official—no matter how self-serving it may be—is to risk getting scooped.

I'm part of the American generation that has made environmental protection a priority in this country, not just a side issue. I take great pride in that. The fundamental change transcends party lines, in fact it transcends politics. It's part of an outlook that holds the future just as important as the present. I thought Linda Maraniss's project was commendable, if not earth shaking. And for a down-ballot official campaigning without much money in a state as big as Texas, there were definite advantages for me. TV crews, maybe even from Houston. Free sound bites. A politician's dream.

As long as I could make it back in time for kickoff.

I grew up inland. But like most Texans, even those who live on the high plains of the Panhandle or in the desert west of the Pecos, I have a special affinity for our coast on the Gulf of Mexico. Part of its lure is our mythology. In 1528 a hurricane pitched the raft of a shipwrecked Spaniard, Cabeza de Vaca, on the Galveston Island beach, and the idea of Texas was born to recorded history. In 1836 Sam Houston coaxed Santa Anna's army of Mexico into the ambush on a lowland coastal prairie called San Jacinto that gave our state its unique heritage as an independent nation. In 1840, at the height of the frontier Indian strife, a Comanche chief named Buffalo Hump led a force of warriors that laid waste to Texas farms and towns all the way to the coast. They rode their horses in the tide-wash as settlers took to Lavaca Bay in boats.

But mostly the appeal is nature, and associations with pleasure. Skies fill up in the fall with flocks of migrating cranes, geese, ducks, traffic jams of hawks, all bound for the Gulf. Five national wildlife refuges are scattered along the Texas coast. One brought whooping cranes back from near extinction; others provide habitat for brown pelicans, sandhill cranes, bald eagles, peregrine falcons. In most parts of the world, dolphins appear in pods of three to seven animals. In the Gulf of Mexico they've been observed in pods of up to 700. Breathing in the thick salt air, you ride a ferry across to a barrier island and watch them rise and dive playfully in the wake, graceful as deer.

Periodically during my childhood, my extended family—cousins, aunts, uncles, sisters, grandparents—would all head for the beach. I remember my dad and my uncles going surf fishing and returning to our motel with that day's catch in buckets and washtubs of ice. I remember making elaborate sand castles and climbing on the playground equipment at Galveston. There weren't many elegant motels or hotels then, in the fifties and early sixties. Often our parents didn't make reservations, but we always found a place to stay. Across the street from the ocean, or sometimes several streets away, depending on the state of the family pocketbook.

It's not uncommon that newcomers to Texas are let down by their first sight of the Gulf. The water doesn't have the clarity and bright color you see along the shores of Florida, and it's hard to get a Californian excited about the surf. Platform oil and gas wells are out there, in plain sight. It's been called the Third Coast and, more rudely, the Redneck Riviera. But then the newcomers unwind and start walking and watch their kids react to it, and pretty soon they're hooked.

The way our beaches are made is a marvel. Rivers flow across Texas's flat coastal plains, leaving gravel in the beds. Grains of sand bounce and roll along, accreting around the river mouths, while silt is flushed out into the bays. Waves wash into the coastline curve at an angle, rearranging the sand and seashells in lateral strips that become our barrier islands. Padre Island, the largest barrier island along U.S. shores, is 113 miles long and no more than two and a half miles wide.

Unlike beaches in most of our coastal states, nobody can buy Texas beaches and fence them off. Thanks to an "open beaches" law that harks back to a time when the beaches were used as wagon roads, and today is the national standard of public access legislation, the only owners are the people of Texas. Gulls hang squawking in the breeze, plovers skim along on their frantic little feet, and the colors of sand and sky merge reflected in the shallows. You scout for sand dollars, watch kids build sand castles, and walk for hours, forgetting how far you've come. It's hypnotizing. Texas has some of the best sand beaches on earth.

I was thinking about that as our plane sat on the runway in Austin that Saturday morning. To direct attention to the cleanup, I had called a press conference earlier that week at the Capitol with two veteran state senators from the coast, Carl Parker of Port Arthur and Carlos Truan of Corpus Christi. We talked about the growing problem of trash on the beaches and agreed that something had to be done about it. Exactly what, of course, was a question none of us could answer.

I watched the rain and the ominous look of the clouds, and I would not have shed any tears if the pilot had told us to forget it, he couldn't take off. But he got us airborne, and we made small talk on the short flight over from Austin. As we circled down to the little airport at Port Aransas, the clouds finally parted, and I peered out the window at the beach below. I was told that we would find the shoreline at very low tide, right after a high tide had receded, which meant there should be lots of debris to work with.

Beer Can City, here I come, I thought. Get those TV cameras ready.

Until I grabbed a trash sack and started working up the beach with Linda Maraniss and the other volunteers, the significance of that combination of low tide and high tide escaped me.

About 2,700 volunteers scoured 122 miles of beach that day. They picked up 124 tons of debris, and a first hand look informed me that the problem had little to do with station wagons and hand-carried coolers, or beer and soda pop cans tossed aside by thoughtless tourists.

We found a lot of hardhats, computer material, and other stuff that was easily traced to the platforms and seismographic vessels of the offshore oil and gas crews. But most of it—about seventy percent—was plastic debris that had been casually dumped overboard from oceangoing ships. For the first time, that could not be disputed. The volunteers filled out computer cards that inventoried everything, even the cigarette butts. The evidence was no longer anecdotal; now there was a database. The inventory identified markings in thirteen languages. Tubes of shampoo from Denmark, toothpaste from Italy, detergent from Singapore. Ancient maritime tradition was turning Texas beaches into a *garbage dump*.

As promised, Linda got me back to College Station in time to watch my beloved Aggies, and I yelled myself hoarse. (Two touchdowns and a field goal behind, they roared back to beat Baylor in a 31-30 thriller that *Texas Football* magazine rated the best Southwest Conference game of the 1980's and one of the best in conference history.) But what I'd seen and experienced that morning on the beach of Mustang Island would not go away.

I was outraged. How in the world had this happened?

That summer the Center for Environmental Education had published a report on pollution in the Gulf of Mexico. I had paid little attention to it. But when I got back to the Land Office I found a copy and started reading.

For 4,000 years, ships under every nation's flag had dumped their garbage into the sea. Prior to the Second World War, the practice didn't create much of a problem. Organic waste is devoured by birds and fish, and most solid matter decomposes in the ocean brine—even aluminum cans—in a matter of weeks. But the invention of plastic quickly made other kinds of containers obsolete, and whether the plastic is buried in landfills or floating in the sea, its resins don't biodegrade for hundreds of years.

I was carrying on about this one day to Tom Henderson, who had recently joined the Land Office staff as a legal counsel. Tom and I have been close friends since we were students at Texas A&M. When I was a sophomore and he was a freshman, he ran my first campaign for Aggie yell leader. We cut our teeth together in politics. And he's one of the brightest policy strategists I've encountered anywhere in government.

In 1971 and 1972 I put off my last year of law school at the University of Texas—okay, my Aggie pedigree is not unblemished—to carry the bags of Ralph

Yarborough in his last campaign for the U.S. Senate. He lost in a Democratic runoff. Later that year Hillary Rodham and Bill Clinton, who were not yet married, recruited Tom and me and became our friends when they interrupted their Yale law studies and moved to Austin to run the Texas presidential campaign of George McGovern. (I know that's supposed to mark us all as wild-eyed radicals, but my answer has always been: Look at the alternative. Richard Nixon was ultimately the one shamed and disgraced, not McGovern. Our candidate just got hammered.) In 1974 I managed the first congressional campaign of Bob Krueger and sent Tom out to west Texas; to everyone's surprise, the Shakespeare-quoting English professor won a seat by capturing the imagination of roughnecks and sheep and goat ranchers. In 1978 we again worked for Krueger when he ran against Senator John Tower and lost. And in 1982 Tom helped me get elected land commissioner at the age of thirty-four.

Does that history make me a career politician? Well, I have stayed involved. But the whole point of politics is having a chance to direct government in a way that helps make people's lives better. Tom and I were schooled in common sense state government by a brilliant, demanding, and often irascible mentor, Bob Bullock, then the state comptroller, now our lieutenant governor. When we were working for Bullock at the comptroller's office in the seventies, he used to say he could hire a hundred lawyers to tell him why he could not overcome a bureaucratic obstacle and make progress. But the men and women he wanted working for him were those who could show him how he could move government forward and make it work for, not against, the people. Bullock's creed has guided us ever since.

"There's not a litter problem on those beaches," I told Tom. "It's an institutional garbage problem. We've had this press conference now, and I don't have any idea what the solution is. I don't want to get us embarrassed. Why don't you look into this and see if there's anything directly that the Land Office can do?"

Tom asked, "Do we even have any authority?"

It was a good question.

2

Lay of the Land

The Texas General Land Office is a unique institution born of our state's unique past. When Texans overthrew Mexican rule and established a frontier nation in the spring of 1836, the legislature created the Land Office by the end of that year. They instructed the land commissioner to collect "all records, books, and papers, in any way appertaining to the lands of the republic, and that may be in the care or possession of all empresarios, political chiefs, alcaldes, commissionarios, or commissioners for issuing land titles."

In other words, the paperwork was an unbelievable mess.

The first land commissioner, a twenty-four-year-old veteran of the Battle of San Jacinto named John Borden, set out to find all the Spanish and Mexican land grants, patents, surveyors' field notes, and other documents authenticating the ownership of over 200 million acres. In a wagon.

Appointed to his post by President Sam Houston, Borden employed a Spanish translator. At one point, he stored the records in friends' homes. When the seat of government was moved from Houston to the new capital of Austin in 1838, he carted 5,000 pounds of paper inland from the coast.

John Borden's brother later founded the famous dairy company that bears the family name. Though seemingly endless land brought settlers like the Bordens to Texas, the agricultural bonanza did not keep our infant nation from going broke. When the government was organized and the first treasury notes were issued, Texas was already $1.25 million in debt. Nine years later, in 1845, the U.S. Congress was disposed to annex Texas but wanted nothing to do with our debt, which by then had ballooned to $10 million. (Because of the slavery issue and the likelihood of war with Mexico, a good number of congressmen wanted nothing to do with us at all.)

So Sam Houston cut a blockbuster of a deal.

When states came into the Union as territories, they turned over their public lands to the federal government. It was just part of the price of admission. That's why almost ninety percent of Nevada, for instance, is today owned by the U.S. government and managed by the Department of the Interior.

In contrast, Sam Houston convinced the federal

government to let Texas keep its public lands so it would have some way of retiring its national debt.

As part of that agreement, he talked the federal government into letting Texas keep the offshore jurisdiction it had enjoyed as a province of Spain and Mexico and then an independent nation. Spanish law established the offshore limits at three marine leagues—everything within a line drawn 10.3 miles from the coast.

Sam Houston's stroke of genius and luck laid the foundation for the economic powerhouse Texas would become. And once the records were straightened out, Houston's vision defined the job of the Texas Land Commissioner. The first state constitution made the land commissioner an elected official, and over the years about 50 million acres have been given to education institutions. But the legislature used land to meet many of its obligations, and for the rest of the nineteenth century, land commissioners presided over a massive transfer of public land into private hands.

Statehood required the surrender of large parts of present-day New Mexico, Colorado, and Wyoming in the so-called Compromise of 1850. (A loss mourned to this day because it included Santa Fe and several top Rocky Mountain ski resorts.) Texas agreed to that in order to retire the republic's debt and because our original Panhandle extended far north of the federal government's established line between slave and free states. But since the national debt was $10 million, and the federal treasury paid Texas $10 million for the lost territory, the transaction cost us nothing but land. In a state that became known for them, Sam Houston was the giant of Texas wheeler-dealers.

The state sold or gave away land to attract settlers and railroads and to reward military veterans for their service—a sense of obligation carried on today by the state's Veterans Land Board, which the land commissioner chairs. In 1873 the legislature set aside half of all the remaining public domain for the public schools. But the privatization of public land continued. In Austin, that period contributed the most famous ex-Land Office employee, a draftsman named William Sidney Porter. He went to work for a bank, got sent to prison for embezzlement, learned to write, and became one of the masters of American short fiction, using the pseudonym O. Henry.

Toward the end, Texas got a new Capitol built in exchange for three million acres in the Panhandle that became the XIT cattle ranch empire. There were only 350,000 people in the state, but their goverment was conducted in the seventh largest building in the world. The new pink granite edifice was up there with Westminister Abbey, the Taj Mahal, and St. Peter's Basilica. By 1898, titles to over 160 million acres had been issued, the Texas Supreme Court declared the public domain exhausted, and in 1900 the legislature passed the Settlement Act to square accounts with the School Fund.

Today—in total contrast to states like Nevada—

more than ninety percent of the land in Texas is privately owned.

In hindsight, one might debate whether pursuing that policy to such an extent was altogether wise. But the fact is, it happened. Knowing and respecting that history goes a long way toward understanding contemporary Texans' reverence for private property rights. And, for that matter, our aloofness toward the federal government.

In a 1919 piece of legislation called the Relinquishment Act, the state sold 7.4 million acres and gave up all but one-sixteenth of the mineral rights to the surface owners. But by then it was known that Texas was sitting on one of the world's great reservoirs of fossil fuels, and in the struggle over oil and gas law the Supreme Court declared the Relinquishment Act unconstitutional—the legislature could not just give away assets of the School Fund. In the court's clarification of the Relinquishment Act, the state retained all the mineral rights, and the surface owners were legally obligated to act as agents for the state in leasing those mineral rights.

When the federal government found out how much oil and gas lay beneath the floor of the Gulf, federal bureaucrats went after Sam Houston's contract with a vengeance, trying to limit Texas's offshore jurisdiction to the three miles held by all other coastal states except Florida (which also benefits along its Gulf Coast from Spanish heritage and legal precedent). But in 1953 Congress confirmed Texas's ownership of the 10.3

mile offshore claims, and the U.S. Supreme Court upheld that legislation in 1960.

The Land Office today manages assets from 20.4 million acres; nearly 20 million of those are subsurface mineral rights with 4.5 million submerged in the Gulf. Apart from the state parks system, the state's remaining surface public domain is concentrated in about 900,000 acres of desert rangeland west of the Pecos River. The state owns the rights to 16,000 producing oil and gas wells. Thanks to oil and gas revenue, the endowment for public education, which is called the Permanent School Fund, is now worth more than $14 billion. And the Permanent University Fund is worth more than $4.8 billion.

So in large part here is what I'm elected and sworn to do as Texas Land Commissioner: I manage the state lands with the mission of maximizing benefits to our school children and college students. That requires detailed knowledge of the oil and gas industry. I'm also responsible for protecting the state's natural resources in the Gulf of Mexico.

I also administer, and during my terms have expanded, the state's loan programs for military veterans, enabling them to buy rural land, make home improvements, and obtain mortgage assistance. That creates jobs and sales in construction and real estate– at no cost to the taxpayers, because the veterans' fees and loan repayments finance the programs.

At the Land Office we have set the standard in state government for equal opportunity employment and contracting. Through beach protection, oil spill prevention, recycling, sustainable energy, and natural gas and clean air initiatives, we have shown that environmental protection can be a boon to the economy.

Education. Economic security. Equal opportunity. And the environment. Like I told those delegates to the Democratic National Convention, it's a great job.

My predecessor was a progressive guy named Bob Armstrong. He's now Assistant Secretary of the U.S. Department of the Interior. Bob was elected land commissioner in 1970, served twelve years, and introduced the idea that conservation of our natural resources and heritage is an essential duty of the office.

He forced energy companies to contractually agree that drilling on state land must be done in an ecologically responsible manner. He also got them to agree to give Texas an incredible 25 percent royalty on new oil and gas wells. Most states get a one-sixth royalty—not quite 17 percent—from wells drilled on their public land, and the federal government only gets a one-eighth or 12.5 percent.

But Bob had to spend all his political capital winning those huge concessions, and important people in big oil, the lobby, and the legislature never forgave him for it. They were always shooting at him, calling him a liberal and, even worse, an environmentalist—in Texas in the seventies that was just short of being called a Communist. And while state government spending was up across the board, they never stopped slashing the Land Office budget.

During his second term I was twenty-eight years old and had gone to work for Bob Bullock in the comptroller's office. Bullock had been telling me about the Land Office and what an important role the Permanent School Fund plays in state government finance.

One night I ran into Bob Armstrong at a party in Austin and told him that if he ever decided not to run for reelection, I sure would like to know—I might like to run for land commissioner myself. It was just party talk. And looking back, I'm surprised he didn't dismiss me as some pushy kid. But five years later, darned if he didn't call.

During that first campaign, I made a speech one day to some high school seniors in Amarillo. If they were eighteen I wanted them registered, I wanted their votes. Of course, a topic that was a lot more important to them than the identity of the next land commissioner was where to go for their senior trip. Port Aransas? Or Chicago? They told me the debate came down to that every year, every senior class. That really struck me and stayed with me. Those kids lived about as close to Chicago as they did to Port Aransas. But even way up in the Panhandle, Texas beaches have this special lure and magic.

After I won the race that November, my old friend Roy Spence and I took our families to South Padre Island for a vacation. Roy is a founding partner of Austin's GSD&M advertising agency, which has consistently astounded the industry by taking accounts away from New York's Madison Avenue. Roy did some political campaign work for me and others in the early years, before discovering that clients like Southwest Airlines, the *Wall Street Journal,* and Wal-Mart pay a lot better, and no doubt sooner, with a fraction of the hassle.

He had also helped launch the most effective anti-litter campaign that the state and, many believe, the country, had ever seen. Roy's firm came up with the slogan "Don't Mess with Texas" for the state transportation department's efforts to clean up our roadsides. The theme was macho and funny, appealing to Texans' pride. Nobody was going to accuse Willie Nelson and George Foreman of getting involved in a nerdish undertaking. And it worked. Even bubbas understand now that throwing beer cans on our highways is not considered cool. In Texas today you hardly ever see the trash-filled bar ditches that I vividly recall from my youth.

But the same could not be said for the beaches. Roy and I took a lot of long walks together on that vacation, and we saw an alarming quantity of trash. I was thinking, "Okay, I got elected—now what?" Roy reminded me of a memorable and effective campaign poster that Bob Armstrong always used. Armstrong defined himself and his role in government by walking on the beach. "When people think of the Land Office," Roy insisted, "they think of beaches. That's the connection they make. Garry, they'll never understand what you do, and they'll never think you're a successful land commissioner, unless you figure out a way to clean up the beaches."

"But there's no way we can do that," I kept saying. "There's no law. There's no money."

Heading into my second term, I was still concerned about sticking my nose in somebody else's business.

Tom Henderson was poring over the statutes, and it appeared that the highway department had the strongest claim to state jurisdiction, under the anti-litter law, because of the tradition of frontier beach roads. But the cities and counties were the only government entities conducting regular cleanups, their crews were swamped by the volume of debris, and they weren't getting any help from anybody except the volunteers of Linda Maraniss and the Center for Environmental Education.

After looking into it, Tom suggested that there was certainly one thing the Land Office could do. The inventory database showed that between ten and fifteen percent of the beach debris was coming from offshore oil and gas platforms and seismic vessels. Nobody could question our agency's responsibility for them. So we drew up an emergency rule, implemented a few months later, that required offshore operators in Texas waters to draw up, post, and enforce a solid waste management plan. The new rule basically said: If we catch you dumping garbage in state waters, you are subject to forfeiting your lease and losing your operating permit.

That's a big hammer, but it was also the only weapon we had. We couldn't just fine the oil and gas operators, because the legislature declined to give the Land Office authority to impose administrative penalties. But with the new rule, we sure got their attention.

An executive of one company wrote us: "Don't you think that lease forfeiture is a little severe?"

We wrote him back that maybe he had a point. We would consider changing our position if he would work with us to get administrative penalty authority from the legislature. We never heard from him again.

We braced ourselves for a furious reaction from the oil and gas industry, but it was mild. Leaders in the energy industry knew they were vulnerable. Climate and geography have given the Gulf something of a break for offshore energy production. We don't have the deep waters and severe winter storms of the North Atlantic, and we don't have the unstable tectonics of the Pacific's ocean floor. Don't get me wrong; we have had our share of oil spills and live in constant fear of them. (More on that later.) We've also had more than our share of pollution by mainland petrochemical sources and ships that flush their tanks at sea. But by and large, the industry's environmental performance here has far exceeded its reputation in New England and California.

Still, industry leaders knew that a lot of Texans resent seeing those platforms in the Gulf. Tar would have been on those beaches in the time of Cabeza de Vaca and the Karankawa Indians because of natural seepage on the ocean floor. But energy executives understood that offshore drillers and producers get blamed every time tourists come back from the beach and have to spend half an hour scraping tarballs off their feet. Requiring employees to responsibly dispose of solid waste was an easy and sensible thing to do.

One query did grab our attention. It came in response to our requirement that the companies post signs about the new rules on all offshore platforms. Someone in the industry said, "Do you really think it's necessary to post the signs on *unmanned* platforms?"

Now that was a realistic complaint.

However effective and reasonable, measures directed at the oil and gas companies could deal with only a fraction of the problem, and it was clear from the start that any comprehensive attempt to clean up Texas's 367-mile coastline would have to be volunteer-based. Neither the Land Office nor any other state agency had a budget for it.

After working out preliminary details with Tom and other Land Office aides, I announced in a series of press conferences in November 1986 a plan to greatly expand the efforts of the Center for Environmental Education's volunteer program, and to make it ongoing and self-sustaining. We called it Texas Adopt-A-Beach.

The overall scheme was a brazen theft from the Texas Department of Highways and Public Transportation. For several years, a foundation of its highly successful anti-litter campaign had been the Adopt-A-Highway program in which church, civic, and other volunteer groups took responsibility for a mile of roadside on the outskirts of each Texas town and cleaned up the right-of-way on a regular basis. The groups were recognized for their efforts with prominent and highly coveted blue and white signs. To a remarkable extent, it kept the roadsides clean.

Many states had followed suit, and so did we. The Land Office and our private sector partners would enlist groups to adopt one mile of Texas beach and conduct cleanups at least three times a year. We also modified the slogan Roy Spence's advertising firm had developed for the highway department's program— the theme on our T-shirts and bumper stickers was "Don't Mess with Texas Beaches." (We gave the highway department full credit for coming up with all these good ideas, and no one ever complained.)

To get Adopt-A-Beach off the ground, we first recruited a task force with a very broad base. City and county officials, representatives of the tourism, oil and gas, and waste management industries, environmental groups, students, and the National Park Service. These people come at the problem from many angles, sometimes with very different and conflicting agendas, but they all had a common stake in the health of the Texas coast.

The group met for the first time in December 1986. One of its members, Dale Bietendorf, then mayor of Port Aransas, said he was there because it was one of the few times he remembered state officials taking an interest in a local problem *after* an election. The task force published program guidelines the following February. The Land Office would sponsor two cleanups a year—one in conjunction with the Center for Environmental Education and one with Keep Texas Beautiful. The adoptive groups would conduct, on their own, another cleanup sometime in the year.

Sounds easy, but folks like Mayor Bietendorf had to be convinced that this wasn't just a political stunt, and officials were concerned that drawing attention to the problem might hurt tourism. (For several years, one

Nueces County commissioner sent his crews out to clean up beaches in his precinct before each Adopt-A-Beach cleanup. I guess he thought the program reflected badly on him, and that the volunteers would eventually get disheartened and just go away.)

Texas has about 182 miles of beaches that are easily accessible. These, of course, were the first priority. Once they were adopted, we could start lining up groups that were willing to take on beaches that could only be reached by boat or four-wheel-drive vehicle. But the process went faster than any of us expected. In April 1987, the first Texas Adopt-A-Beach cleanup attracted 3,600 volunteers. The following September, the number leaped to 7,100. They were just ordinary Texans, giving up a Saturday morning to do something that made them feel good about a place they loved. In the process, they had a good time, a weekend at the beach.

Corporations make valuable contributions. Mobil helped get the first cleanup running by providing 10,000 garbage bags. Browning Ferris Industries (BFI) provided trucks to handle the waste collection. The workforce of Dow Chemical has become one of our biggest and most committed groups of volunteers. School groups from as far away as San Antonio and Austin join in the cleanups, and students throughout the state now participate in an annual poster contest emphasizing clean beaches. Lucky the Dolphin, a porpoise that had been found tangled and near death in a synthetic net in the Gulf, became the program mascot and for years was a top attraction at Galveston's Sea-Arama. Every mile of public Texas beach was long ago adopted, and we have a waiting list of groups that would like to participate.

The same approach can be applied to other coastlines, of course, and lake shores and riverbanks as well. Over the years, forty U.S. states and territories and seventy nations have modeled programs after the Texas plan. Costa Rica was one of the first to follow our lead because of a friendship between our first Adopt-a-Beach director, Ingrid Kavanaugh, and the nation's Foreign Minister.

Ably directed at the Land Office over the years since by Angela de la Garza, Jeannie Collins, Susan Ghertner, and Mary Gordon Spence, Texas Adopt-A-Beach recently observed its tenth anniversary. Over that decade, more than 184,000 volunteers have cleaned up 3,709 tons of garbage off our beaches. The problem in the Gulf has not gone away. But along with sparing local governments about $14 million a year in beach cleanup costs, those volunteers represent a constituency that could not go ignored in boardrooms and government forums where the root causes of the problem really can be addressed.

In 1987, the first year of the program's existence, Texas Adopt-A-Beach received the Chevron Conservation Award from Robert Redford and the Take Pride in America Award for volunteerism from President Ronald Reagan. For me it was an honor to meet those men. I have photographs of the ceremonies on my walls. But my role in Texas Adopt-A-Beach was that of a facilitator. Those thousands of Texans filling up garbage bags, working hard enough to make themselves sore, marking the kinds of trash on their inventory cards—they were the ones really being honored.

We live in a time when cynicism too often holds sway. Countless politicians and elected officials are staking what they hope will be long careers on the proposition that American government—the most inspired and envied system of government the world has even seen—is by nature wasteful, intrusive, and bad. They're free to believe whatever they want to believe, but what a strange way to spend their time.

If I was not convinced that our government can be a positive force, that it can be made to work in ways that benefit the people, I would have found other things to do a long time ago. Government can be a catalyst for change and for focusing attention on a problem. Throwing money at a problem is seldom the answer, but providing some resources and organizational assistance often allows others to make real, significant contributions. There is no better example than Texas Adopt-a-Beach. When government makes sense, doesn't squander money, doesn't create a new layer of bureaucracy, and tries to solve problems in ways that harness our best instincts, time and again, the people respond.

3

Underlapping Authority

Soon after I got back from that first 1986 cleanup in Port Aransas, I learned that a national dispute over ship dumping of garbage and plastics had actually been going on for quite some time. The issue then focused primarily on endangerment of wildlife. Waterfowl get tangled up in the plastic, and marine life mistake it for food. A couple of years earlier, an infant pygmy whale had washed ashore on Galveston Island and died from a severe stomach infection. In its stomach, an autopsy revealed, were a plastic bread wrapper, a Fritos bag, and a thirty-gallon garbage can liner.

That's terrible, of course. But politically, endangerment of wildlife added up to a very narrow constituency. Certainly that constituency could be broadened, and it was. The Audubon Society held its national convention in Corpus Christi as the issue was crys-tallizing in Texas, and its members were appalled when they went to see the whooping cranes and found them wading around in plastic debris at the Aransas National Wildlife Refuge. Their outrage got other environmental groups involved, which in turn brought representatives of the hotel-motel industry and other tourism officials into the process.

But few Texans had ever seen a seal or sea otter in the wild, much less taken a swim with one. In California the argument about marine entanglement was being framed as an animal rights issue. I seriously doubted that enough Texans would get exercised about an issue called "marine entanglement" to moti-vate our congressional delegation to lead a charge on Capitol Hill. I'm a firm believer in going directly to the source of the problem. If the source in this case was international shipping, that was the logical place to begin.

So I called a shipper and asked, "Why do you dump garbage overboard?"

"Because the ports don't have reception facilities," he said.

So I called someone at one of the Texas port author-ities.

"Why don't you have reception facilities?"

"Because the U.S. Department of Agriculture has these regulations that won't allow us to receive foreign garbage. It's prohibited. So we don't receive garbage."

This is why people hate bureaucracy.

Even if enlightened members of the shipping industry were disposed to change a destructive practice going back 4,000 years, an agency of our federal government was effectively telling them they had to go on dumping in the sea.

I called an old friend and political ally, Congressman Kika de la Garza of Edinburg. In 1961, when he was a member of the state legislature, he co-authored legislation that oversaw commercial use of Texas's tidelands. His expertise and involvement in coastal issues spanned more than three decades, and in 1986 his congressional district took in one of Texans' best-loved stretches of beach and coastal resorts, South Padre Island. It also included the thriving farming country of the lower Rio Grande Valley. Kika was then chairman of the U.S. House Agriculture Committee.

"Mr. Chairman," I asked him, "why does the USDA impose regulations that mean ships dump garbage that winds up on our beaches?"

"You got me," Kika replied with wry humor and tongue firmly in cheek. "But it's an election year. Let's have a hearing!"

A little over a week before the general election, Congressmen de la Garza and Solomon Ortiz of Corpus Christi conducted a hearing in Corpus that began to put the issue in much sharper focus. I testified. Linda Maraniss testified for the Center for Environmental Education. A lot of essential players attended those sessions: representatives of Keep Texas

Beautiful, the city of Corpus Christi, the port authorities of Houston, Corpus Christi, and Brownsville, offshore oil operators, the American Institute of Merchant Shipping, the Navy, the Coast Guard, and, of course, the U.S. Department of Agriculture.

A man named Bert Hawkins was the administrator of the USDA's Animal and Plant Health Inspection Service. He told us that the regulations arose out of concern that bugs and germs could come into this country with foreign garbage and potentially devastate agricultural production—not to mention threaten human health. When ships came into port, the USDA essentially told the shipping companies and the captains, "We're not going to let you offload your garbage, and if we find garbage on board that is not properly contained, we're going to fine you for it."

The agriculture department bureaucrats' concerns about exotic infestation were entirely valid, of course. But it seemed hard for them to carry the logic further: If you were a ship captain and knew that your garbage might well get you and your employer fined by government inspectors in United States ports—and the easy fix had been standard practice since Phoenicians ruled the seas—what would you do with it?

You'd dump it overboard.

And arguably, the regulations were not even serving the purpose for which they'd been drawn up. As it was plainly evident on the Texas coast, jettisoned ship garbage floats. Some of the bugs and germs ended up getting in anyway. (At the Land Office we learned that, according to one respected theory, fire ants—exotic pests from Latin America that plague southern states and whose invasion is checked only by northern winters and the barricade of desert in the southwest—may have established a beachhead on the Gulf in just that manner.)

In the weeks to come, Tom Henderson and I spent a lot of time making our case in the Washington, D.C., offices of Bert Hawkins and other administrators of the USDA. They were bright and dedicated professionals; it was just hard for them to acknowledge that the regulations might have some flaws, and that their agency was part of the problem. Kika de la Garza reminded them frequently of his personal interest—and that of the powerful House committee he chaired—and in time the agriculture officials started making suggestions on how to find a solution.

And unlike bureaucrats of the U.S. Navy—who from the outset took the position that it was their God-given right to dump their garbage in the ocean—at least the agriculture officials voiced some concern about the problem. At the Corpus Christi hearing, Bert Hawkins summed up the dilemma. Ocean dumping of solid waste was an international practice. It was not against the law. And though it was clearly having a detrimental effect on Texas beaches, not one government agency—local, state, or federal—could say: We have jurisdiction, and here's what we're going to do about it.

He called the problem "underlapping authority."

In other words, trash was falling through a lot of government cracks.

At the Corpus Christi hearing, Joe Angelo, the Coast Guard representative, called us aside and told us there was an unratified treaty we ought to know about.

That was the first time I heard of MARPOL.

In Washington nobody stops to smell the roses. Tom and I were just part of the frantic crowd—suits, briefcases, everyone in a hurry, ten minutes late. The only thing that set us apart was his cowboy boots. That November, one of the times when Tom and I were up there pestering USDA officials, Joe Angelo set up a meeting with his Coast Guard superior, Admiral J. William Kime, the service's chief environmental officer. Kime also headed the United States delegation to the International Maritime Organization.

After introducing some other members of his staff, Admiral Kime told us that MARPOL was actually a package of treaties that resulted from the massive *Torrey Canyon* and *Amoco Cadiz* oil spills from ocean going tankers in the sixties and seventies. (The name was short for "marine pollution.") A 1973 treaty called the International Convention for the Prevention of Pollution from Ships, and a clarifying 1978 protocol which went into effect in 1983 dealt exclusively with oil spills. In separate compacts, called annexes, the MARPOL agreements on pollution and liability also dealt with hazardous chemicals shipped in bulk, packaged hazardous waste, sewage, and garbage. Annex II, the one covering ocean spills of bulk hazardous waste, would take effect in 1987.

But countries bound by the first two agreements were not required to ratify the others, and it seemed that Annex V, which covered garbage dumping at sea, was going nowhere fast. In order for the annexes to join the body of international maritime law, they had to be ratified by at least fifteen countries representing fifty percent or more of the world's shipping tonnage. More than twenty countries had ratified Annex V, but their combined shipping only added up to about forty percent of the world's tonnage. The United States-flagged fleet, to our surprise, accounted for just under five percent of the world's shipping tonnage, but its influence far exceeded that modest figure. The U.S. Senate had not ratified Annex V, and from our discussion it sounded like the State Department had it stewing on a far back burner.

Admiral Kime told us that Annex V would outlaw dumping of plastics in the ocean anywhere in the world. It would prohibit discharge of other floating garbage within twenty-five miles of a coastline. And it would require port authorities to provide reception facilities.

Another provision, he said, designated "special areas" where no dumping of anything except food waste was allowed. If the treaty was ever ratified, the Red, Black, Baltic, and Mediterranean seas and the Persian and Oman gulfs would be granted this additional protection. These were small, enclosed or semi-enclosed basins with little circulation, and were exposed to a lot of shipping traffic.

"That sounds like the Gulf of Mexico," I said. "Why isn't the Gulf designated a special area?"

The silence went on so long that it scared Henderson. We were always concerned about revealing our ignorance. Admiral Kime, who was to become one of the best friends that Texas beaches ever had, looked at his aides, and they looked at him. He said, "Garry, nobody ever asked me that question. I don't know. "

"Well, could we make it one?"

"Maybe. And that's a great idea. I'm going to help you."

The Land Office staff and I were bewildered, exhilarated, driven by sheer possibility. How many times would we have a chance to help move the enormous weight of government toward change that was clearly beneficial and had global significance? Of course, we knew it was also an opportunity to demonstrate that we were in completely over our heads, playing way out of our league.

All at once we had three agendas.

How could Texans influence the Senate Foreign Relations Committee, which was consumed at the time with South Africa and the Iran–Contra hearings, to focus on an obscure treaty dealing with garbage?

Assuming MARPOL Annex V could be ratified, what was the best way to implement and enforce it?

And what would it take to convince the International Maritime Organization to designate the Gulf of Mexico as a special area worthy of extra protection?

No treaty was going to stop irresponsible individuals from dumping garbage in the sea. Pick any law and somebody's going to break it. But the structure of international agreements provided a way to fight an institutional practice. As an expedience—as a business practice—the maritime institutions were taking a position and assuming a right that, instinctively, no ordinary American citizen believed in. That made it very hard for the shippers to defend themselves.

But if the Land Office was going to play any role, first we had to make our case in Texas. Our plan was to keep it simple—define this thing as an issue of beaches. Drawing on the intense public interest that Texas Adopt-A-Beach had tapped into, I started making speeches wherever I could. I talked to county commissioners, local parks officials, Rotary and Kiwanis clubs, anyone who would listen.

Very few people are unaware that Florida is only ninety miles from Cuba. But unless you look closely at a map, you may fail to notice that the western tip of Cuba is almost that close to Mexico's Yucatan Peninsula. Except for those two narrow channels, the Gulf of Mexico is a relatively shallow, landlocked sea.

Hydrology and topography create a series of looping currents that move water in the Gulf like it's in a bathtub with a slow drain. Driven by wind and tides, longshore currents hit the curving Texas coastline at oblique angles. Those currents are one of the natural forces that lay down the strips of sand and shell that become our fine beaches. But they also send a disproportionate amount of the Gulf's flotsam our way.

I had a big map that emphasized the virtual land bridge and detailed the currents. But for clarity, it was hard to beat Tom Henderson's phrasing. "The ship garbage is just out there floating," he'd drawl, "and the coast of Texas and northern Mexico is kind of like a big catcher's mitt."

Wherever I went, the reaction was the same: incredulous. You mean it's not against the law to dump garbage in the ocean?

We enlisted groups like the Western States Land Commissioners Association and the Coastal States Organization, and worked hard to spread the word and line up support in other Gulf coast states. We started helping Louisiana organize a volunteer beach cleanup, and during the winter and spring Tom and I made one or two trips to Washington a month. We found that the Coast Guard had been communicating with the groups that were focused on marine entanglement, but that was about all the discussion going on.

So we started talking our way into meetings with the Audubon Society, the National Wildlife Federation, the Environmental Defense Fund, the Sierra Club—and the Oceanic Society, to whom they generally deferred on issues like this. We learned that because of budget considerations environmental groups in Washington have very specific agendas. And it was obvious from the skewed glances they gave us that they were not used to sharing much time with people in state government, particularly from Texas. They were sure we must have had some other agenda.

On every trip, we also called on members of the Texas congressional delegation. One time it would be Austin's Jake Pickle, the crusty Democrat who held the seat that, before him, sent Lyndon Johnson on his path to the presidency. On another trip it would be Mickey Leland of Houston, who had been my friend and inspiration and who used to saunter into the Texas legislature sporting a dashiki and an Afro that would have filled up most of a bushel basket.

On other occasions we went to see Sugarland Republican Tom DeLay, whose district bends from west Houston to the coastal bottomlands around Lake Jackson. Tom, a former pest exterminator who served several terms in the Texas Legislature, was then just starting his second term in Congress. He's now the House majority whip, the third-ranking leader of the Newt Gingrich brigade, and some of his environmental views are well-known and commented upon frequently. Global warming is a fraud. The bald eagle was never endangered. DDT is perfectly safe. And so on.

But you know what? I always got along with Tom DeLay and often found that I could work with him. When he was in the legislature and I was in my first term as land commissioner, he helped me on agency matters several times. And now, as I explained what brought us to Washington, he got all excited and said, "This is great! First of all, I can do something about a problem in my district, and second, I may be able to cast my first environmental vote since I've been up here."

4

Plastic, Rice, and Red Beans

In early December 1987, with Tom Henderson an official member of the U.S. delegation, Admiral Kime first floated the idea of special area designation for the Gulf of Mexico at the meeting of the International Maritime Organization in London. The response was encouraging. Mexico's representative said he couldn't react without instructions from his government, but on its face the proposal had merit. Even more promising was the reaction of the Europeans, who are extremely influential in the maritime community. The delegates of West Germany and the Netherlands spoke in favor of Admiral Kime's proposal, and not just because they were concerned about the Gulf. Enlarging the list of designated areas made terrific sense to them. Scandinavian and other countries got excited about special area designation protection for the North Sea—which of course helped us.

On January 21, 1987, President Reagan sent Annex V of the treaty to the Senate for advice and consent on ratification. Senator Lloyd Bentsen voiced his strong interest in the issue, and though he was not a member of the Foreign Relations Committee, he began to use his considerable influence to move ratification higher on the agenda. On the other side of the Hill we gained another important ally—Congressman Gerry Studds of Massachusetts. Congressman Studds chaired a House subcommittee on fisheries, wildlife conservation, and the environment, and he wanted to get a bill on implementation of Annex V moving through the House.

Not everyone was on board, of course. The Associated Shrimpers of New Orleans didn't want anyone telling them what to do with their harvest nets, their chicken bones, their red beans and rice. Nothing. In Houston the president of the West Gulf Coast Maritime Association was furious at me; he said I was laying far too much blame on the merchant marine fleet. He didn't like our "numbers," which were drawn from the beach cleanup's computer database. But he acknowledged that the shippers were part of the problem, and when asked to supply industry statistics for use in my speeches, he said he didn't have any numbers at all. The Port of Houston Authority feared it would lose business to ports on the Atlantic and Pacific if the treaty was ratified and the Gulf was designated a special area. I told all of them that if there had to be a fight, it was going to be in the press, and in the legislature, and they were going to lose this one. Because I knew none of them would be willing to go public and say, "We stand on our rights as spoilers of the beaches and seas."

The exception to that was the Navy. Nobody fought MARPOL Annex V as hard as the United States Navy.

At the congressional hearing that Solomon Ortiz and Kika de la Garza had called in Corpus Christi, the Navy ship environmental program manager stated flatly, "We do not believe that Annex V coverage should apply to Navy vessels in a strict sense." Whatever was required of private ships, the department's imposition of a twenty-five mile offshore no-dumping zone was the best the Navy could do.

The program manager went on to say that Navy regulations required that its sailors attempt to package the trash, but often this was impractical. But after many years of research and development, a special trash compactor was almost ready for shipboard test and evaluation; fleet-wide installation could begin in about four more years.

Storage of garbage onboard would eliminate space for weapons systems. It would create fire, safety, and sanitary hazards for the crew. Plastics were the way of modern life. Sailors couldn't be expected to separate out their cleansers, razors, and film canisters from degradable garbage.

Plastic trash bags seemed to be the worst problem, he thought, "in both the operational and aesthetic sense." And yet they, too, were indispensable. "We would encourage industry to develop a biodegradable plastic bag. . . . We feel so strongly about the need for biodegradable bags that we are planning a small research effort this year to produce such a bag for our use. That concludes my statement."

In other words, forget it.

At the time I was related by marriage to a retired admiral in the Navy. Following the Corpus Christi hearing, he kindly referred me to another admiral, who declined to speak to me directly. Instead, on one of those first trips to Washington, I should meet with some of his subordinates.

At the Navy shipyard in Washington, drab warehouses stretched as far as I could see, I was dealing with guys who had thirty-year-old desks. Navy captains. They were shipping out soon to some other assignment, and they let me know very clearly that ship garbage was way down the list of their priorities. If there was anything lower, it was me. In time we insinuated our way a little farther up the chain of command. At one point the Navy officials started talking about a multi-year study that would cost nine million dollars and would perhaps result in a thermonuclear waste incinerator. They weren't kidding!

Mayor Dale Bietendorf of Port Aransas had become one of our best allies in Texas. Dale listened to the spiel in one of those meetings and said, "You guys ever heard of Sears and Roebuck? They got some trash compactors that are pretty small. Fit right under the kitchen cabinet. I bet you could put some of them on ships and they'd just about get the job done."

Congressman Studds was pushing a House resolution on implementation, and in June he invited me to testify at a hearing in Washington. In the lobby of the House office building that morning, my aides and I confounded the security personnel. After running our briefcases through the metal detector, they told us to empty a suspicious-looking canvas bag. Rolled out before them was our traveling exhibit of plastic tubes, bottles, squeeze containers, hanks of yarn.

"What is this?" asked one the guards, reluctant to touch it.

"Marine debris," someone said.

"What?"

"Ship garbage. It's a long story."

The stuff made a great prop. Before the subcommittee, I pointed out that of 73,613 foreign ship landings in the United States in the prior year, only 1,731 resulted in offloading of garbage. In Texas the ratio was worse—20,000 to 90. Ratification of Annex V and designation of the Gulf as a special area were essential steps in the process of improving that record. But we had to have an effective way to enforce the measure.

"The shipping industry advocates onboard incineration," I told the House members. "You could provide incentives to install that equipment. Port facilities must be made capable of receiving offloaded garbage. The USDA supervises garbage disposal from international flights at all international airports—ordinarily through contracts with private waste–management companies. Why can't a simple, private-sector, cost-efficient program deal with ship garbage in the same manner?

"If ships can't burn their garbage, they should be required to present and offload it, just like we do at international airports. If they don't, they should be denied port entry or fined. Make presentation and offloading a component of a solid waste management plan required of every ship. If ships have crossed the ocean and have no garbage to present or offload, we can all guess what they did with their garbage; we know what they did with it. It's in our seas and oceans.

"Our current situation is similar to the one we faced in the 1960s when cities and industries bled open sewers into our streams, rivers, and lakes. First, we had to close the sewers. Technical fine-tuning came later in the process. Advocating the Clean Water Act twenty years ago, another Texan, President Lyndon Johnson, said, 'There is no excuse—and we should call a spade a spade—for chemical companies and oil refineries using our major rivers as pipelines for toxic wastes.' Nor is there any excuse today for the world's fleet using our shores and our submerged lands as a garbage dump."

I didn't think the language was all that inflammatory. But the technical fine-tuning soon had the Texas General Land Office and its commissioner in the middle of a firestorm. And the heat was not coming from foreigners.

The Coast Guard, for all its sympathy, understanding, and motivation, complained about the budget increases that would be needed to patrol the Gulf and other coastal waters within the twenty-five mile limit. In their minds, they would essentially be writing the ships' captains tickets. The Coast Guard could fine them $25,000, which was nothing to sneeze at. But how were they ever going to catch anybody?

Well, if the dumping was done at night, Coast Guard helicopters had these infrared detection devices

That didn't sound realistic. It seemed to us there were only three possibilities: You burned ship garbage, you offloaded it, or you dumped it overboard. And to us that was the logical premise of an enforcement strategy. Make the shipping companies come up with a solid waste management plan—just like the offshore oil and gas producers in Texas coastal waters—and demonstrate proof to U.S. port authorities that they were implementing it.

A couple of weeks after that subcommittee hearing, Tom Henderson represented the Land Office at a big meeting in Washington. Gerry Studds wanted to start marking up the bill. There were about forty people in all—from the Coast Guard, the USDA, the Environmental Protection Agency, the merchant marine fleet, the passenger ship lines, and congressional offices on both sides of the Hill.

Tom was floored when the subcommittee staffer

opened the meeting, thanked everyone for coming, and then turned the meeting over to him. Tom talked about our experience in Texas, about the sensible concessions made by the offshore oil and gas producers, but when he got to presentation and offloading, representatives of the port authorities, the shipping industry, and the stevedores threw a fit.

This plan was an outrage. Why, it was going to cost millions and millions of dollars. Why should they have to pay for it?

A former New Hampshire congressman turned lobbyist for the cruise ship lines was particularly incensed. Here was an industry whose lifeblood was tourism, and whose public image was shaped by the feel-good TV series *Love Boat*. But its lobbyist knew that vacation ships were one of the biggest sources of floating glass and plastic in the sea.

"Congressman," said Tom, "do you throw your garbage in the street behind your house?"

"Of course not."

"Do you throw it out the window when you drive down the street?"

"Of course not."

"Then why do you think you ought to be able to throw it off the back of your boat?"

The meeting went on like that for two or three hours.

Finally the representative of the merchant shippers said: "Well, it seems to me that making everybody offload at every port is ridiculous."

"Okay," said Tom. "What's a better idea?"

"In most cases, they're going to be able to store

garbage for several days. Why don't you have them enter on the captain's log every day what they do with it—and show where and when they've offloaded?"

That's often how the legislative process works. Someone comes to the table completely opposed to the entire proposition, and in the give-and-take makes a knowledgeable remark that reveals the core of a consensus. The captain's log. On the high seas it's almost holy writ. Captains who get caught falsifying their logs can have their licenses revoked. A very simple record-keeping procedure became the centerpiece of an enforcement strategy that everyone could work with. Everyone except the Navy.

The Pentagon took the position that it could not accommodate any treaty that prevented garbage dumping at sea because finding the space would make them downsize crews or eliminate weapons systems. This was a matter of national security.

At a reception on the Hill one night, Congressmen Kika de la Garza and Jake Pickle confronted the Secretary of the Navy, James Webb, and told him that the brass was feeding him a load of poppycock. Not long after that, Tom Henderson got on an elevator in the Longworth office building and realized he had walked into a heated argument. Kika was going at it with a fellow Democrat, Charles Bennett of Florida, who chaired a defense subcommittee on naval affairs. Congressman Bennett walked with two canes, but that didn't slow him down one bit. He was animated and angry: "The Navy tells me that if they have to deal with this garbage stuff, they're either going to have to remove weapons or men."

"Charlie," Kika shot back, "it's just obvious that you've been brainwashed."

"Well, if you're so smart," Bennett shouted, "why don't you run the Navy?"

He burst out the door and stormed down the hall, canes just flying.

But the national security argument always carries weight; the Pentagon had Gerry Studds' implementation bill stalled. The proposed bill granted the Navy a phase-in grace period of five years. The Pentagon claimed that even that was impossible. The earliest the fleet could comply was in fifteen to twenty-two years.

Twenty-two years? This was 1987. These were the same folks who were telling the nation they could put up the Star Wars missile defense system in three!

Kika de la Garza couldn't believe it. "Well, I'm an old Navy man, but this is ridiculous."

Neither could Jake Pickle. "Well, I'm an old Navy man, but this is absurd."

The Pentagon's arrogance aroused and unified the Texas delegation in a way that no environmental issue could. Marvin Leath, who later left his central Texas seat to pursue a career as a country singer, growled: "Those Navy guys are gonna make a liberal out of me yet." Within forty-eight hours, Kika, Jake, and Henderson were running around the Hill collecting signatures for a letter to secretary Webb from the Texas delegation asking for acceptance of the five-year phase-in. "The Navy's estimated time frame of fifteen to twenty years is unreasonable," it said.

The only members of the Texas delegation who didn't sign the letter were Senator Phil Gramm and Congressman Solomon Ortiz. (The letter moved too fast to suit Senator Gramm, or at least his staff, and Congressman Ortiz was under considerable pressure from the Navy, which was considering Corpus Christi for a "Home Port.") But the letter bore the signatures of Senator Lloyd Bentsen and Speaker of the House Jim Wright. House Democrats Mickey Leland, John Bryant, Martin Frost, Ron Coleman, and Henry B. Gonzalez. Republicans Dick Armey, Tom DeLay, Bill Archer, Joe Barton, and Jack Fields. On many issues, this was about as partisan and divided a bunch as you could find.

When Congressman Studds received a copy of the letter during a committee hearing, he interrupted the testimony of a Navy officer and announced in his precise New England accent: "I have just been handed a letter signed by the entire Texas delegation, which as you may know is not without influence in this House . . ."

The Navy yielded, knowing it was just looking silly. For us it had been an amazing thing to observe and be involved in. I always thought that if I could find the one Navy decision-maker responsible for ship garbage, and just have a conversation, reason would prevail. I never did find that decision-maker. Still haven't, to this day. I don't think that person exists.

5

Cuban Cigars

Newcomers to the Texas coast are often startled to see people driving on the beaches. Texans have been using the hard–packed sand between the tidewash and the vegetation line for roads since they traveled by horse and wagon. When the state was being settled, it was impossible to build permanent coastal roads on the mainland because of the soggy marshes. So the travelers crossed over by ferry to the barrier islands and made their way up and down the coast on the beaches. The beach roads show up on the first maps of the Texas highway system. And that history of unbroken public use is the reason that the only owners of those beaches today are all the people of Texas.

Since humans first took to the seas, they've used

beaches to land and regroup—to repair ships, sails, and fishing nets. Roman and English common law established that tradition as a public right. Of course, that hasn't stopped the rich and powerful in this country from buying up beachfront property and telling the public they can't set foot on their private beaches. But not in Texas.

One of the most gifted legislators our state has ever produced was a Houston congressman named Bob Eckhardt. During his eleven terms in the U.S. House, Eckhardt helped define the workplace safety standards of the Occupational Safety and Health Administration. Defying Lyndon Johnson, he was one of the first congressmen to speak out against the Vietnam War. Carrying on for Ralph Yarborough, in 1974 he engineeered final passage of the bill creating the Big Thicket National Preserve. In 1979 he argued passionately and with eventual results that if the federal government had any conscience, it had to compensate Nevada cancer victims who were exposed to fallout from eighty-six atomic bomb tests. And, with the aid of a young Tennessee colleague named Al Gore, he fought off attempts to gut the Superfund program of toxic waste cleanups.

But before Eckhardt went to Congress, he served in the Texas House of Representatives, and it was there in 1958, as a freshman, that he designed and engineered passage of the Open Beaches Act. Many think it was the most impressive work of his long career.

Developers were determined to carve Texas beaches into the enclaves of profit and privilege found in other states, and the first thing they tried was to put up obstructions to vehicle traffic. Eckhardt knew that there was no way the state could afford to keep the beaches open through condemnations—even if the votes for that could be found in the legislature. He argued that Spanish law and frontier use of the beaches as coastal roads constituted a presumption of public ownership of the beaches up to the vegetation line. It was a stroke of legislative genius.

Galveston's A.R. "Babe" Schwartz, who arrived in the legislature two years after the act was passed, carried on grandly in the sixties and seventies after Eckhardt went to Washington. Chairing a Senate committee to study the beaches, Babe Schwartz steamrolled legislators, lobbyists, and lawyers with orations about the Public Trust Doctrine and the Institutes of Justinian, a sixth–century emperor of Rome. "Those rights are not granted by the Texas Legislature," he said, shaming them. "They've been our rights since time immemorial."

Jim Mattox, a former Dallas legislator and congressman who was elected Texas Attorney General in 1982, defended the public beaches all the way to the U.S. Supreme Court. Mattox was never known to back off from anything, but that was a particularly courageous stand to take. In 1983 Hurricane Alicia moved the vegetation line back behind some homes on Galveston Island. Politics in Texas always favors private property rights. But allowing those homeowners to rebuild would destroy the premise of unbroken public use just as surely as putting up traffic barricades.

The court agreed with Mattox that the public's rights to the beaches were indeed paramount. The Open Beaches Act has held up against every challenge, and because of it, Texas has the best beaches in the country and they belong to ordinary Texans, their children, and generations to come.

In June 1987 I described the Texas law at a congressional hearing on beach access in Rhode Island and again the next day in a speech at the University of Rhode Island in Narragansett. New England has some of the least accessible beaches in the country, and a lot of residents resent that. At the time, the region also had an unusually large number of United States Senators on the Committee on Foreign Relations—Clairborne Pell, John Kerry, Chris Dodd, George Mitchell. I talked about beach access but also used the Rhode Island forum to press for ratification of MARPOL Annex V. This was a beach issue, too. Write your senators, I urged people, and a lot of them did.

That spring and summer a New York barge called the *Break of Dawn* had been cruising around the Atlantic and the Gulf of Mexico, trying to find someplace to dispose of 3,100 tons of garbage. It was one of those comic misadventures that captures people's attention and sheds light on the larger, underlying problem. All at once garbage in the ocean was an issue—a story—and it started to get a lot of play in the press.

We got a huge break: The Houston bureau of *The New York Times* put together a piece about MARPOL that ran on the first page, and when that happened, it was like the floodgates opened. Within twenty-four hours we had calls from the major television networks and *Newsweek* and *Time*. We steered the reporters to Admiral Kime, Congressman Studds, and the Oceanic Society. Admiral Kime urged ratification of the treaty at a Senate hearing in New Jersey in July. Meanwhile, Congressman Studds' implementation bill was being marked up in the House, and there's no question that the September letter from the Texas delegation to the Navy got the attention of the Congress. In October the House approved the implementation bill by a vote of 386–14. And that month, the Senate Foreign Relations Committee sent the treaty to the floor.

The next break was even bigger. In the course of that debate, Senator Bentsen, whose support had been unwavering and crucial, called a press conference and said he was going to carry an amendment that, as part of the ratification process, would put the Senate on record in favor of special area designation for the Gulf. Senator Phil Gramm and other colleagues in Gulf Coast states agreed to co–sponsor the amendment.

Admiral Kime had told us that the International Maritime Organization would require technical evidence of the need for special area designation, but that the Coast Guard did not have the funds to pay for such a study. So we found the money, and scientists engaged by the Center for Environmental Education supplied the report, which was circulating that fall.

MARPOL is basically an agreement signed by the world's primary shipping countries. Because of financial requirements, a lot of developing nations are not signatories. But there are other ways to accomplish the same purpose. One of those is a regional seas agreement called the Cartagena Convention, under which countries in the Caribbean could enact a protocol that would parallel Annex V's special area designation. That fall Caribbean delegates to the Cartagena Convention were meeting on the island of Guada-loupe in the French West Indies.

Mexico and Cuba, whose eventual support was crucial, have coastlines on the Caribbean Sea and a lot of influence in the region, and there were compelling reasons besides politics to broaden the scope of special protection. From the Venezuelan mouth of the Orinoco River, currents in the Caribbean send debris up along the shores of Central America to the curve of the Yucatan Peninsula and into the Gulf through the straits of Mexico and Cuba.

A friend in the State Department called us at the Land Office and said, "You need to make sure you have somebody in Guadaloupe. We've gotten instructions from headquarters that we are not to raise the issue of garbage. That's coming directly from the Department of Defense. The Navy is really hot about that letter from the Texas delegation. But if you guys raise the issue, then we can help."

So Henderson, after securing the required approval of Governor Bill Clements' budget director, was off to the Caribbean, where the Colombians soon jokingly referred to him as Mr. Ambassador from the Republic of Texas.

The Senate ratified the treaty on November 5, 1987. The White House circulated the treaty for comment and review by the Coast Guard, the Department of Transportation, the National Security Council, and the State Department. President Reagan signed the documents for transmittal to the International Maritime Organization, but the State Department took the position that it couldn't proceed until Congress finalized implementation legislation, which was being held up by a jurisdictional fight between two committees in the Senate. To shake it loose, across the Hill Congressman Studds attached the bill to an important fishing rights agreement between the U.S. and Japan, which had to be finalized by the end of the year.

Finally it looked like a done deal. Just a matter of administrative procedure. The U.S. would be able to present the treaty to the IMO at its next meeting in London in February 1988. And all at once, that was urgent: More than enough nations had ratified the treaty, but it could not become law until the signatories' shipping tonnage added up to more than half the world's total. The U.S. and the Soviet Union got in a race and public relations contest to see who could put the treaty over the top. But during the Christmas holidays, somewhere between the bureaucracies and the White House, the treaty got lost.

Literally. Nobody could find the papers.

What's more, it seemed that nobody had alerted the White House that the documents were on the way. Henderson and I were called back into the fray in a way that made us squirm. In 1978, as I mentioned before, he and I had worked for Bob Krueger in his campaign to unseat Senator John Tower. It was a hard-fought race that turned exceptionally bitter; from Houston, the press wire carried a famous and startling photograph of Tower refusing to shake Krueger's hand. As Krueger's campaign manager, I was the designated controversialist. My candidate lost that race by less than 10,000 votes. It was one of the lowest points in my life. Now, nine years later, the congressional liaison in the Reagan White House was a man named Will Ball—John Tower's former top legislative aide.

Of course, Will Ball knew exactly who we were. But he was a professional—and a Texan—and never blinked an eye. He got the White House staff prepared to move quickly on the treaty, and he helped us find the missing documents. And that's how politics should work—used to work. Once the campaign's over, you may not always forgive and forget. But public policy is too important to the people government is supposed to serve to be held hostage in a partisan grudge match.

On December 30, 1987, just fifteen months after Linda Maraniss's troop of volunteers spent a Saturday in Texas picking up trash, wearing T-shirts that said "Be a Beach Buddy," U.S. documents ratifying the treaty arrived at IMO headquarters in London. We beat the Russians. And one year later, MARPOL Annex V became international law.

Special area designation for the Gulf of Mexico took a good while longer. Following an extensive search to track down the scientist who originally compiled the study that became the basis of the MARPOL treaties, we learned that he had always intended that the Gulf would be on the list of special areas. It obviously met all the criteria. But those agreements were directed primarily at shipping spills of oil and hazardous waste. The big oil companies were so paranoid about possible restrictions of traffic in the Gulf that they had insisted that the special area designation be taken out. In order to gain support for the treaty as a whole, the United States went along with dropping the Gulf from the list.

Admiral Kime was promoted to another post, and the admiral who succeeded him as the Coast Guard's environmental chief was not inclined to push Annex V any further. Among the mossbacks there was a theory that if the Gulf became a no-dumping area, that would give Caribbean ports the excuse to impose dockage fees to pay for dealing with the garbage, and that would restrict the flow of navigation. A sizable contingent of Coast Guard officers and bureaucrats subscribed to that view.

It certainly helped that two Texans from the coast, George Bush and James Baker, were now President and Secretary of State. That got a lot of calls returned. On the other hand, the shipping industries, especially the cruise lines, had the ear of the Vice President, Dan Quayle, and his Council on Competitiveness. For almost two years, I thought that special area designation was dead.

The balance tipped back suddenly when Admiral Kime got another promotion—to Commandant of the Coast Guard, its top officer. The admiral in charge of environmental affairs might have been unenthusiastic about special area designation, but now his superior told him he had to be for it, so he was.

Admiral Kime contacted the Mexicans and started talking to them. If they opposed special area designation, it would be very hard for other Central American countries to go along with us. Also, Mexico has diplomatic relations with Cuba, and Admiral Kime knew that Mexico was the key to getting the Cubans on board, which in turn was key to winning IMO approval.

But I didn't think we'd get much help from Mexico's foreign ministry. In the seventies I spent twenty months as special counsel to the Mexican desk at the State Department. With all due respect, I came away convinced that Mexican bureaucrats make American bureaucrats look good.

We had seen some very interesting photographs that convinced us that Cuba's beach garbage problem rivaled ours in Texas. How could we pursue that commonality of interest? In Congress, Houston's Mickey Leland was always trying to open lines of communication between the United States and Cuba. In the 1962 missile crisis, Fidel Castro did in fact help push the world to the brink of nuclear war, and he has brutally hung on to the Western Hemisphere's longest–standing dictatorship. But Mickey's deepest concerns as a politician were always hunger, poverty, literacy, and health. Mickey's perspective was that whatever else could be said about Castro, he has delivered universal health care and universal education to his nation's poorest of the poor.

Mickey had a relationship with the Cubans, whether he was supposed to or not, and it was managed by a congressional aide named Luis Delgado. In August 1989, Mickey had died in an tragic plane crash while on a hunger relief mission to Ethiopia. Luis had left Congress for the private sector. I had known Luis for years, and I hired him as a Land Office consultant to approach the Cubans on the subject of special area designation. Luis went to Cuba and talked to Cuban officials in Washington (where their interests were managed out of the Czech Embassy). Luis took Henderson and other Land Office staff to a meeting there later, and then to New York to talk to the Cuban ambassador to the U.N. A couple of days earlier, this Cuban had been seen on national TV shaking his fist in the face of Secretary of State Baker over matters in Iraq.

Nobody said a word or even registered a facial expression as Tom spoke at length, perspiring. Then suddenly the atmosphere became all relaxed and pleasant, and the ambassador got out a bottle of rum. He broke the seal and started twisting the cap, but it wouldn't come off. "The Russians make our caps," he told our guys, laughing. "It's always that way with the Russians. Round and round. Nothing happens."

We kept quiet about our contacts with the Cubans, especially to the Coast Guard, but our friends at the State Department knew what we were doing. There was a lot of back-channel communication. And meanwhile, Secretary Baker was doing all he could to get the Mexicans to go along. At an annual meeting between the nations' cabinet-level officers, he put special area designation on his personal agenda and argued for it forcefully.

With Mexico and Cuba on board, the next task was trying to make sure none of the nations bordering the Gulf of Mexico and the Wider Caribbean—as the special area was now known—would object at the upcoming IMO meeting. Working with the Coast Guard, the IMO's Caribbean regional office, and a U.N. environmental programs office in Jamaica, we helped set up a forum for discussion among those nations at an October meeting in Caracas, Venezuela, concerning a regional pollution protocol.

Henderson was now a member of the U.S. delegation to Caracas. After some animated discussion, the twenty-six countries in attendance agreed not to oppose the special area designation—all that was required for IMO approval—if help could be secured from the World Bank and other international financial entities to install proper reception facilities.

After the election in 1990, Admiral Kime invited me to be a voting member of the U.S. delegation to the meeting of the IMO in late November. Henderson by then had worn out his welcome with the Coast Guard admiral in charge of environnmental affairs. The admiral was still ambivalent about special area designation at best, and the sight of Tom had him all but grinding his teeth. Especially when Tom not only arrived in London, but with credentials as a voting delegate of the International Friends of the Earth.

It was not a glamorous trip. The IMO is housed in a very impressive building overlooking the Thames, but all I could do was glance out the window now and then and think how pretty the city was. The sessions went from early in the morning till late at night. I never saw my hotel in daylight hours. I'd walk instead of taking a cab just to get some sense of London.

The ambivalent admiral kept telling us that this was going to take at least two to five years. The Cubans would just never agree. He didn't know we already had an understanding with the Cubans, along with the Mexicans and Venezuelans, whose young representative, Emma Toledo, was placed in charge of the committee to draft the proposal—the first woman in IMO history to attain such standing.

Still, it almost fell apart. The bureaucrats struck again. The Coast Guard had all these technical reasons why the language wouldn't work. The admiral made some remark that Emma Toledo thought was patronizing. Henderson and the admiral were screaming at each other. It came down to the last day, and still we couldn't agree on the language. I was playing good cop to Henderson's bad cop, and finally the Coast Guard agreed to the wording we thought we had to have. The admiral assured us there was no chance the Cubans would ever go along. But when we presented the document, the Cubans initialed it on the spot. It was a total surprise to the admiral. There was nothing he could do but sign the paper, too.

It was the first time the United States and Cuba had officially agreed on anything since the airplane high-jacking accord of the 1970s.

A few weeks after that, I was back in Washington on other Land Office business, and the Cuban attaché invited me to breakfast. I could hardly turn him down. I picked the most public place I could think of—the middle of the Hyatt Regency—because everyone knew he had FBI and intelligence agents following him around. The whole time I kept my hands on the table.

I smoke a cigar or two a day. I kept thinking, He's going to give me a box of cigars, I've never had a Cuban cigar. And he did. Monte Cristos, I hoped. But the cigars were a brand I'd never heard of.

I put the box in my suitcase, and when I got back to Austin, I stuck it in a drawer. I was a little indignant. If it was a brand I'd never heard of, it couldn't be any good. But then somebody gave me a book on the history of cigars, and I read that during the revolution, one of the guys who went up in the sierra with Castro rolled his own cigars, and Castro loved them. After the war he helped the guy set up his own company—Cohibas. Put away in my drawer for several months had been a box of Castro's favorite brand of cigars. You can only do so much for the boys from Bryan.

Under the terms of MARPOL Annex V, the Gulf of Mexico was designated a special area in July, 1991. There should never have been a doubt that it would get additional protection if the other enclosed and heavily navigated seas and gulfs did. But for twenty years bureaucrats and big oil blocked it.

Today the framework of law outlaws dumping of anything except food waste anywhere in the Gulf. It requires ship captains to keep track in their logs of what they do with garbage, just as they're required to do with oil and hazardous waste. Because it recognizes that there are only three things you can do with ship garbage—burn it, offload it, or dump it overboard—the law is very easy to enforce. And Texas beaches are cleaner because of it.

As Texas Land Commissioner, did I ever think I would be a delegate to the International Maritime Organization? No. Did I ever think I'd ever be negotiating with the Cubans? No. But we were willing to think unconventionally, to go outside the box, to do whatever it took to accomplish the objective. If it hadn't been for the Land Office—and my dedicated, innovative staff—I am certain that none of this would have happened. Many of the allies and friends we made in Washington are still pursuing careers in diplomacy and foreign policy, and they tell us that because of the fight for the treaty and special area designation, the State Department learned that on some matters, involving the states can be extremely useful in furthering American interests. Above all, it showed that the enthusiasm of average Texans could overcome even the inertia of federal bureaucrats and the international maritime community. It was a triumph of Texas common sense.

6

Grandma's Rubber Bands

As a boy, when I walked into my grandmother's house on Fairway Street in Bryan, I knew that if I looked behind the front door, I'd find dozens of rubber bands looped around the doorknob.

Every day when she went out and got the morning paper, on the way back inside she'd roll off the rubber band and put it on the doorknob. She figured sooner or later she'd need them. And I bet she never bought a rubber band in her life.

She was the same way with ribbon and wrapping paper. When she opened Christmas or birthday gifts in her house, she did it very carefully. She picked off the Scotch tape then neatly folded up the ribbons and paper and put them away in a box to use again on presents she gave to us. She was the same way with aluminum foil. In the fourth drawer of her kitchen cabinet were squares of foil she'd saved from bake sales at church. That was one thing she wouldn't have to spend money on. And she never lacked for aluminum foil.

Today that's called "recycling and reuse." I think of my grandmother when I hear about some poll or read a press account which concludes that recycling is a trendy fad. People aren't prepared to spend one dime more for recycled products, the argument goes. They're not going to do anything real for the environment, so they ease their guilt by carrying out their papers, cans, and bottles for curbside collection.

Baloney. They're just doing what used to come natural. I'd be surprised if my grandmother ever said the word "environment." In her mind, those were just frugal, obvious, and practical things to do. Recycling was no fad to people who lived through the hard times of the Depression and the rationing of World War II. Children wore hand-me-down clothes that had been outgrown by their brothers and sisters, and it was nothing to be ashamed of.

But in the half-century of prosperity since the war we've become a disposable society, often to a ridiculous extent. In Austin two friends of mine have a kitchen dishwasher that just chugs along. Never breaks down. But they've had this dishwasher for fifteen years, and the trays, where you stack the dishes and cups, started showing some wear and tear.

They didn't need a new dishwasher; they just needed to replace the trays. And though the appliance was old, it was a major brand whose manufacturer is still going strong. So they looked in the Yellow Pages and found a business that supplies parts of dishwashers. They got the model number and wrote down the measurements and asked someone at the business to order them the trays.

After looking into it, the man at the parts supply store was embarrassed. Yes, the company still made the parts, and would fill the order. But the upper dishwasher tray cost $170. The lower tray, because it had no moving parts, cost $120. On the other hand, a brand-new dishwasher of the same size was priced just under $300.

Somebody at a corporate management level made that decision. The simple part my friends needed might contain ten bucks in raw materials. But for them it would be less time and trouble, and cost the same, if they bought a new dishwasher instead of keeping the perfectly dependable one they had. Is that nuts?

I know. Anecdotal evidence is unreliable. And it's not that simple. Cost of labor and shipping. Storing. Assembly line space. All that. But what were my friends supposed to do with the old dishwasher?

That's easy: Have it hauled to the dump.

Every time I hear windy discussions about over-crowded landfills and the "infrastructure of recy-cling," I think about my grandmother's rubber bands on that doorknob. My grandmother never read the *New York Times Magazine*, which caused a stir recent-ly by debunking the effectiveness and the intellectual premise of recyling, but nobody would have had to explain the logic to her. She did it because it made sense. It helped make ends meet.

Some of the strongest and most effective environmental initiatives in this country have grown out of that kind of rural and small-town commonsense perception, especially in the Midwest. Look at Oklahoma, which hardly has a reputation for save-the-planet crusades. People there got fed up with seeing used tires thrown in every ditch and gully, and told their legislators they would go along with paying a two dollar surcharge on every new tire purchased, if that's what it took to get the state to properly dispose of old tires. They came up with a way to solve the problem because it was an obvious blight on land they loved. Ordinary people see conspicuous consumption and don't feel good about it. They hear about planned obsolescence and know it's a get-rich scam. Recycling is a matter of going back to our roots. Where I come from, you don't throw away things that still have value.

At the Land Office, our involvement in recycling was a direct outgrowth of the Adopt-A-Beach cleanups. Solid waste problem? Well, three times a year we have several tons of one. When we first inventoried the beach garbage, I was puzzled that we hardly ever found any aluminum cans. But it was no mystery. Every morning on every beach in Texas, someone was out there picking up the beer and soda pop cans. If the beachcombers picked them up, they could sell them. There was a market for recycled aluminum.

Coors beer has had a recycling operation since the 1960s. When the directive first came down from corporate headquarters in Colorado, I imagine a lot of distributors in Texas thought it was a public relations gimmick, if not a pain in the neck. But today recycling is a proven money-maker for Coors.

In order for recycling to work, there has to be a market-driven demand for the products. You have to close the loop. Paper, for example, is one of the easiest things to recycle. Production of recycled paper uses half the water, half the energy, results in far less pollution, and of course it conserves trees, if not whole forests. And eighty percent of the American demand for paper is in private business and industry.

But how could we get on a high horse about the private sector when it turned out that the Stephen F. Austin building, where the Land Office and other state agencies are housed, was trucking out five tons of waste paper a week?

And it wasn't just that we were throwing away paper in our office building. We were mailing paper that people all over the state would ultimately throw away. In the Veterans Land Board, our computers sent out thousands of monthly bills to the veterans repaying their loans. I started pestering the state purchasing agency, the General Services Commission, to supply us with recycled computer paper. They replied that in the first place, they couldn't find competitively priced recycling paper. And even if it was competitively priced, our demands would use up the entire world supply of recycled computer paper. No question, that was an obstacle.

About that time, because of my friendship with Hillary Clinton, I got appointed to Wal-Mart's environmental advisory board, which she chaired. At the first meeting we started talking about Wal-Mart using recycled paper. Bill Fields, who was then the president of the company and the management representative on our board, said that's something they ought to do. At the next meeting, they said they had three companies bidding to provide Wal-Mart with recycled computer paper, and they were going to save money using it. Well, suddenly state purchasing bureaucrats in Austin could no longer say there was an insufficient supply of recycled computer paper and it cost too much.

Another thing Wal-Mart did was to put out the word that it wanted to start selling recycled motor oil. Until then, the oil companies were distinctly disinterested. But they were aware how many people shop in Wal-Mart. Overnight, their attitude changed. The truth of the matter is, none of that has anything to do with the Texas General Land Office. For me it was a revelation on how the market drives recycling.

In every way possible, we started using recycled products at the Land Office. It was amazing to me how many recycled products we could find that were competitively priced. And once suppliers knew we were interested and would consistently buy the product, they were a more than willing supplier. We knew that as demand increased, so would the supply. If those of us in government really believed in recycling, then it was our responsibility to lead the way.

We initiated a recycling program throughout the Capitol complex of office buildings, and in the 1991 session of the legislature, we worked with Senator Carl Parker of Port Arthur and Representative Robert Saunders of La Grange on a bill that committed the state to a comprehensive policy of waste reduction and recycling. Senator Parker's legislation emphasized some of the most dangerous and pressing waste disposal issues—lead batteries, used motor oil, and unwanted tires. The strategy included public education and market development components, and it required government entities to give preference in purchasing to products made from recycled materials. Initially, in the bidding process it allowed them to pay ten percent more for those products. That last provision was phased out after five years on the assumption that as recycling industries proliferated, economies of scale would bring the cost of their products into competitive balance. And for the most part, that assumption has been borne out.

We also organized the Texas Corporate Recycling Council. The stated goals were to establish a preference for recycled products in private sector purchasing, promote the manufacturing of safe products, emphasize recyclable packaging, reduce the amount of packaging, and create new markets for recycled goods. But in actual practice, what we discovered was that when individuals in corporate America made a commitment to recycle, they made money. The Texas Corporate Recycling Council is all about replicating individual success stories: If somebody's making money using recycling technologies in Dallas, why not Houston?

In Northeast Texas, the little town of Big Sandy had built the first commercial composting plant in the country in 1973. They knew that sixty percent of the material going into landfills could be turned into humus that enriches the soil. A company called Vital Earth Resources, which operated Big Sandy's composting plant, took in sludge and other organic waste materials from the region of Tyler, Texarkana, and Longview. The organic waste was processed and cured until it was transformed into some of the highest quality potting soil, compost, and mulching material available in area garden centers.

In West Texas, the biggest employer in Kimble County is a Junction company called Advanced Environmental Recycling Technologies. Since 1985 the company has processed more than 15,000 tons of recycled material and secured ten U.S. patents. The materials they work with are recycled plastic and cedar sawdust. Together they make weather-resistant decking material and window and door frames. Plastic and juniper! It's hard to think of two substances that are more plentiful and hard to get rid of.

For me, one of the most gratifying contributions was being made by the Chief Auto Parts stores. When I was a nineteen-year-old college sophomore, I bought my first car. I was on a tight budget, and when it came time to change the oil, I used a brand of recycled oil called Double Eagle because it was much cheaper. But when I drained the dirty oil out of the crankcase, I would carry the pan out to the alley behind my parents' house and pour the oil on the ground around the garbage cans, so it would kill the weeds, and I wouldn't have to trim them.

Of course, I wasn't the only one who wasn't thinking the whole thing through. A quart of used motor oil can contaminate a year's supply of drinking water for fifty people, and every year in this country, backyard and carport mechanics improperly dispose of enough oil to equal thirty-five *Exxon Valdez* oil spills.

Now I found that Chief Auto Parts had devised a highly successful program for recycling used oil that customers brought in. It returned to the shelves for sale as high-grade motor oil. And they took that responsible policy a hard step farther. Most quarts of motor oil are packaged these days in plastic, not cans. One kind of plastic container is contaminated by the oil and can never be recycled. Chief kept working with its plastics suppliers until they developed a container that could be recycled into other products. All it required was commitment. Closing the loop.

Just about everyone I know who went to college in the 1960s saw and loved the movie *The Graduate*. In the cocktail party scene, the guy tells Dustin Hoffman his one word of advice for the future: "Plastics." Listen, I believe that. Texas is the world's capital of plastics. The plastics industry is worth more than $15 billion a year to the Texas economy. Texas manufactures fifty percent of all the plastic produced in the world. Our state needs those jobs, and in global competition, we need to protect our market share.

But, again, my introduction to recycling as a public policy issue was the debris in those first beach cleanups. More than seventy percent of the garbage the volunteers picked up was plastic. As time went on, I learned that discarded plastic takes up about thirty percent of the volume in our landfills—and that much of the plastic manufactured today will still be around in four hundred years. I also learned that not all plastics are created equal. There are seven major kinds of plastics, and another fifty or so minor types. And at the time, it took an expert to tell which one was which. Recycling plastic is no easy matter—even if people are disposed to try.

In 1989 I was invited to make a speech to the Society of the Plastics Industry. I told the members that their industry should invest at least as much effort and resources into developing recycling technology, and finding markets for products of that technology, as they spent developing and marketing products made from virgin plastic resins. The audience was polite, but I quickly found out that was not a message they wanted to hear. Most leaders in the plastics industry saw reycling as a threat.

The industry had created a couple of pseudo-environmental fronts, but mostly what they wanted to do was produce television spots about how helpful and beneficial plastics are. A couple of companies were marketing "bio-degradable" garbage bags that, to the embarrassment of the industry, turned out to be nothing of the sort. But plastics is a cyclical industry. When profits go up in plastics, the companies always build their production capacity until supply outstrips demand. So the price of plastics drops dramatically. When the Land Office really started promoting plastics recycling, prices were down and there was too much production capacity. So people in the industry responded like this: Let me get this straight. We're losing money, we're selling plastics below the cost of production, and you want us to recycle stuff, in order to reduce demand for our product?

Government consistently fails when it tries to mandate fundamental change in the marketplace. But what it can do is prod and facilitate the change. In 1989 legislation was enacted to require that all plastic containers manufactured in Texas must be stamped with one of seven codes: six denoting the most common resin types and one for mixed or layered resins. An expert is no longer required to sort through the plastic and send it on its way to an appropriate recycler. I always take a container along when I talk to grade school kids, and they love it. Recycling is no theoretical notion to them. It's something beneficial that you don't have to be an adult to do.

And in the marketplace, plastics recycling is no longer a far-fetched impracticality. The Texas recycling bill directed the Land Office to conduct an ongoing market development study for recycled products. At the end of 1992, seventy-one Texas firms accepted some form of waste plastics. Two years later, the number had grown to 111, and twenty-two were manufacturers. Demand has grown substantially for recyclable plastics, especially those made from the PET resins. Recycling of PET containers jumped by 200 million pounds in this country in the first years of this decade; the recycled resin is being used for denim fabric, ceiling pads, even the bumpers on new Ford Tauruses.

But the plastics recycling industry is still in its infancy. Curbside collecting is problematic. For municipalities, collection and sorting is expensive, because it's labor intensive, and the return from selling the plastic is fairly low. For manufacturers, containers retrieved in that fashion tend to be highly contaminated. Transportation regulations make it cheaper to ship plastics interstate than intrastate. And plastics engineers still have to solve the problem of compatibility— compatibility not just of resins, but of colorings, caps, and labels.

But market swings can help recycled products just as they can hurt them. In the spring of 1994 a shortage of ethylene caused the price of virgin plastic resins to jump seventy to eighty percent. In direct response, the price paid for recycled plastic made of HDPE resins leaped two to fourteen cents a pound. And a definite change of attitude can be observed in the established plastics industry. It has no choice but to respond when Chief Auto Parts demands a recyclable oil container. Or when Heinz 57 makes an executive level decision to put ketchup on the shelves in an easily recyclable squeeze container.

Environmental programs that work have one common thread. People are willing to do the right thing, but they have to believe it will make a difference. In the Adopt-A-Beach cleanups, volunteers understood that a month or so later, the beaches would be littered with debris again. But they knew that by focusing attention on the problem, they were giving others leverage to stop the institutional practice of dumping garbage at sea. They didn't want to just pick up bleach bottles from all over the world; they wanted to stop seeing them on Texas beaches, and to a large extent they have. It used to be that when sailors heard the signal of "second bells" on the night watch, that was time to dump the garbage overboard. The Texas beach cleanups demonstrated broad public support for ratification of a treaty that outlawed that practice.

It works the same way with recycling. People aren't going to carry out household waste for curbside collection if they're convinced it's pointless. But they'll do it when they push their baskets through their supermarkets and see competitively priced products made from recycled material. They see the loop closing when they pick up plastic debris in a beach cleanup, check off the items on their inventory cards, watch the garbage hauled off by the trucks of BFI, and then six months later they come back and find that Exxon Chemical, Conoco, and DuPont have turned the waste plastic into park benches that are set up on their beaches. Texans understand recycling.

7

Get the Bugs Out

Easter weekend in March 1989, I was getting out of the shower and had my little TV on as I dried off. The morning news came on about the *Exxon Valdez* tanker wreck in Alaska and the oil spill that devastated some of the most beautiful and sensitive natural areas on earth. I was just mesmerized by the story, and I kept thinking: I am on some task force or committee that's supposed to deal with oil spills in Texas. I don't think I've even been to a meeting. If an oil spill like that happens in our coastal waters, it's going to be all over the Texas beaches. Not only am I going to get blamed; I ought to. I have no defense.

Oil spill response authority in Texas was fragmented at best and non-existent at worst. The Texas Water Commission had the statutory responsibility for spill cleanup, but it had no practical knowledge of coastal oil spills. The Railroad Commission exercised authority for spill cleanup only when oil and gas exploration or production facilities were involved. You couldn't hold anyone accountable, and no one felt accountable.

When I got back to the Land Office that week I asked two people on my staff, Dudley Lightsey and John Riley, to make themselves experts on oil spills. "Let's find out everything there is to know about oil spills," I told them. "Let's ask all the questions, go to every meeting."

They learned that on upland oil spills in Texas, both the Railroad Commission and the Texas Water Commission use backhoe crews who build earthen dams around the oil. Then they sort things out and decide who's responsible and who has to pay for getting the mess cleaned up. Both agencies were treating marine spills that way, too. Unfortunately, there's no way to dam the ocean. You have to deal with an oil spill in an ocean environment the same way you respond to a fire. The equipment had better be in place, and you had better be there immediately. You can't wait for days.

The Alaskan Pipeline ends at the Port of Valdez, and on paper, it had the best conceived oil spill plan in the country. It was the most controversial pipeline in history, and was over-engineered to address environmental concerns. But when the oil companies that built it and used the port began to downsize, the first thing on the chopping block was environmental safety. The cuts eliminated entirely some of the equipment needed to execute the plan and rendered other equipment inoperable. It was Easter weekend, and many of the personnel responsible for executing the plan were on vacation. You know the result.

After the catastrophe in Alaska, it became clear that Texas was woefully unprepared to deal with a major spill on its coast. I went over to the capitol with testimony that I thought was a real call to action: "But for the grace of God, Texas could be enduring what Alaska's suffering today." I pointed out that sixty percent of the nation's imported oil enters the country through Texas ports. "It's not a question of 'if' we have something resembling an *Exxon Valdez* in Texas. It's a question of 'when.' And we are not prepared to deal with it."

When I finished, not one legislator asked a follow-up question. No reporter ever filed a story. It was like I was invisible—hadn't said anything. I didn't expect them to pat me on the back for being concerned. But I did think this was a fairly glaring policy flaw in state government. Nobody cared. I was stunned.

Texas ports receive about 4.6 billion gallons of oil every year. Yet, while other coastal states made major overhauls of their emergency response plans, no bills were introduced in the Texas Legislature. Our only reaction to Alaska's horrendous experience was a minor revision of the weak and diffuse law already on the books.

Legislators weren't the only ones who ignored the problem. Governor Bill Clements commented that there was no way the state could fully prepare for a major oil spill. Of course, this was the same man who said environmental damage from the massive IXTOC oil spill a decade earlier was "much ado about nothing" and said the solution to the problem was to have a good hurricane.

Following the conclusion of the 1989 legislative session, I was appointed to serve on an oil spill advisory committee. This was no lightweight committee: also serving were the chair of the Texas Water Commission, Buck Wynne; the chair of the Railroad Commission of Texas, Kent Hance; the chair of the Texas Parks and Wildlife Commission, Chuck Nash; and Robert Lansford, director of the Governor's Office of Emergency Management. The results of these committee meetings left much to be desired. The committee insisted on issuing a report to preserve the status quo and keep oil spill response authority fragmented. I went on to issue a minority report that called for a coordinated program built around dedicated personnel and pre-positioned equipment strategically placed along the Texas coast.

One day Peggy Sackett, who was my secretary when I had a private law practice, came by and said she had a friend who had a process that could help solve the problem.

"By all means," I said, "send him over."

One afternoon, a tall man came to the Land Office and introduced himself as Carl Oppenheimer. He wore glasses and was in his sixties. He had a little plastic bag that contained an orangish-yellow substance. He also had a little bottle of oil. I have several aquariums in my office. Dr. Oppenheimer walked over to one of the tanks, opened the bottle of oil, and tilted it over the clear blue water. "May I do this?" he asked politely.

"Yeah," I finally said. But I immediately thought he was going to kill my fish.

He didn't seem worried as the oil dripped into the water. Then he opened the plastic bag and poured the orangish-yellow substance into the tank. Half an hour later the oil was gone, and the fish were eating the residue that was left. He said that microorganisms had turned the oil into fatty acids, which the fish evidently found quite tasty.

Now I'm no scientist, but I've learned to believe my eyes.

Days went by. And I didn't lose one fish. There was no oil in my aquarium.

That was my introduction to the promise of bioremediation in fighting oil spills. I was aware, though, that microbes had been used for decades to neutralize and clean up toxic pollution. In fact, by getting elected land commissioner I was "manager" of one such application.

Back in the sixties, in eastern Harris County, near the San Jacinto River, a chemical company named French Limited had used a gravel pit as a toxic waste dump. The French Limited gravel pit and dump became one of the most dangerous plots of terrain in Texas. The buried contamination was so bad that it became a federal Superfund site. Only the costliest and most nightmarish hazardous waste sites qualify for that federal assistance.

When John Hill was Texas Attorney General, his office had sued the chemical company, and forced it into bankruptcy. Well, some assistant attorney general decided it was a good idea to seize the company's assets—and then dedicate them to the School Fund! So some of the most contaminated real estate in Texas was transformed into state land. Thanks a lot, fellas.

I knew that bioremediation was one of the techniques being applied to the French Limited cleanup by the EPA, which conducts the Superfund program.

Carl Oppenheimer also claimed to have strong evidence that the same sort of biotechnology could be useful in responding to oil spills. To tell this story, I need to credit Galveston journalist Tom Curtis, who traced it back to its origins. In the early fifties a railroad fireman and self-taught West Texas scientist named James Francis Martin invented something he called "the living water." It was derived from seawater, cow manure, and yeast.

Fermentation turned this liquid into a substance that stimulated rapid reproduction of microbes that literally ate contaminants of water and soil, and it could do other useful things—such as enhance secondary recovery of oil deep in the ground. James Francis Martin understood the vast implications of what he had discovered, but nobody paid much attention to him, and he died virtually unknown, and no doubt frustrated, in 1975.

Carl Oppenheimer was a professor of marine science and microbiology at the University of Texas in Austin. He had two Fulbright fellowships on his résumé. Dr. Oppenheimer had been trying to find a way to put microbes to use in oil spills since 1969, when President Nixon appointed him to a panel studying the Chevron blowout off the coast of Santa Barbara, California. Bill Blakemore, a Midland oilman and rancher who had served as the chair of the Texas Department of Public Safety, knew about James Francis Martin's discovery. In 1984, through Mr. Blakemore's efforts, Dr. Oppenheimer finally had the catalyst for which he had been searching.

In 1985, the scientist, the oilman-rancher, and other associates started a company called Alpha Environmental. Their clients were looking for oil in problematic subsurface structures in the U.S. and the Soviet Union. And Dr. Oppenheimer was waiting for a chance to test the bugs in an oil spill.

He had gone all over the world collecting samples of aggressive oil-eating microbes. He said he had specimens from more than eighty sources. In October 1989, five months after the *Exxon Valdez* spill in Alaska, with the cooperation of the EPA, Alpha Environmental, and Ruska Laboratories of Houston, we started testing his microorganisms at the University of Texas Marine Science Center in Port Aransas. We filled three big concrete tanks with water, and put fish, oysters, and shrimp in all three tanks. We poured oil in two, and left one as the control tank. In the tank treated with microbes, the result was the same as it had been in my office. The marine life survived, and the oil appeared to have vanished. All that was left was a yellowish crud which the fish quickly ate. Later, Carl Oppenheimer collected the fish, oysters, and shrimp and ate them for dinner. By scientific standards, the results were inconclusive. Academics debunked our report, of course. Our methodology was not rigorous enough. But then I was just an Aggie trying to find a way to get oil off the water.

Why was the Land Office conducting bioremediation research? Because hardly anybody else was. Academic scientists in the field had been encouraged to choose another specialty and find another way to make a living. Twenty years earlier, as a cost-cutting measure, the Carter administration had eliminated the funding for research and development of oil spill technology.

It didn't take long for my dire predictions to come true. On June 8, 1990, a Norwegian supertanker called the *Mega*

Borg was transferring or "lightering" oil to the Italian tanker *Fraqura*. The Mega Borg exploded and caught fire, spilling an estimated 4.6 million gallons of light Angolan crude oil into the Gulf fifty-seven miles from Galveston.

Four crew members died in the explosion; seventeen others were injured. It was one of the worst spills in U.S. history.

The 885-foot tanker had been carrying about 38 million gallons of oil and was "lightering" or transferring some of its cargo to the smaller Italian tanker when the blast occurred. It was touch-and-go for a few hours whether the ruptured ship would sink with all that oil. As bad as the *Exxon Valdez* spill had been, in comparison it put eleven million gallons in the water. The ferocious fire off the Texas coast burned for a week and was hot enough that it almost melted the ship's steel superstructure. Once the tanker was fully ablaze, crews had to spend a week slowly cooling it down while its cargo of crude poured into the Gulf.

When a tanker is offloading oil, fumes are a real problem. Elaborate safety measures are put in effect on all tankers, but they're computerized. No noses are sniffing to determine whether vapors are building up. Evidently, the *Mega Borg* had a computer failure, then a vapor buildup and then a fire. Then the ship exploded, and there were at least several hundred thousand gallons of oil in the water.

Although much of the spilled oil either evaporated or burned in the fire, we were concerned that the slick would reach Texas beaches, particularly at Galveston, and the thick marshland near Sabine Pass. We were also concerned about two national wildlife refuges in the area—Anahuac on High Island and J. D. Murphee near Sabine Pass—and the Sea Rim State Park. On the Louisiana side of the border, the Sabine National Wildlife Refuge was vulnerable.

I went to Galveston and started trying to find out what was going on. When I arrived, I attended a meeting consisting of the Coast Guard, state and federal officials, county officials, and a representative of the responsible party. I discovered that due to the nature of the high seas and position of the spill, booms were useless. The response "plan" was to allow as much of the oil as possible to evaporate or burn and for the rest of the oil to wash up on the beaches which were to act as cheap boom.

We then learned through the on-scene coordinator that NOAA's computer program indicated that the oil would hit the beaches within the next forty-eight hours. I thought, "Well, at least we have forty-eight hours to boom the beaches."

I continued to ask the question: "What are we going to do to protect the beaches?"

The Coast Guard officers said nothing could be done for the beaches. All they could do was try to protect the channels leading into Galveston Bay.

"What about booms?" I persisted.

"There aren't any."

"Can't we fly them in from somewhere?"

"We could start making them now and have them ready in five or six days."

I knew how much clean beaches mean to the local economy. I knew that this was *no* plan for Texas, and so did all of the local folks involved. How could I or anyone else explain to the people of Galveston that federal bureaucrats were going to use their beaches for cheap boom.

Then and there I vowed the beaches of Texas would never again be used for boom.

The next day we learned the NOAA model we had relied on had been wrong. Instead, oil was expected to hit the beaches in two hours or less. The Land Office then began recruiting volunteers to help with the cleanup when the oil arrived. Volunteers arrived in droves and were mobilized by Land Office staffer Blanton Moore.

Fortunately, the federal bureaucrats were wrong again. The oil never washed ashore, but the *Mega Borg* was a wake-up call for Texas. It was beneficial in another respect, as well. We had fought the waves and oil, and with luck had won. Yet another, and in some senses, bigger fight was taking place.

When I was first informed of the *Mega Borg* spill, I immediately saw a real-world opportunity to put Carl Oppenheimer's microbes to work. I jumped at the chance to use the spill as a field test, but faced immense resistance from the Coast Guard and academic bureaucrats from other federal and state agencies.

I had strongly urged the on-scene coordinator and other federal bureaucrats to use bioremediation. It was truly an uphill battle when you consider that the EPA did not bother to send a representative at first, and when he did arrive, a decision could not be made until his superior flew all the way in from Washington.

The real tragedy was that all I wanted to do was use a product and a technology which had already been approved by EPA. Chemical dispersants were also on the same approval list and their use had been approved for this spill. If dispersants could be used, why not bioremediation?

Fed up with bureaucrats, I knew the only way I could get approval for bioremediation would be to get the people of Texas on my side. I needed the media to help tell my story, and I had never seen this many TV cameras in one place. This was big news around the world, and cameras had come from as far as Russia to cover the story. CNN was covering the event live. I had fish tanks brought in, called a news conference, and repeated the same experiment which had worked in my office and at the UT facility at Port Aransas. Again, the oil dissolved into fatty acids and the fish lapped it up. This time on TV.

Then I told the Coast Guard that unless they used bioremediation, I would call another press conference. So we got the approval. They told us there was a cutter leaving Freeport at 4 a.m. and to "have your bugs there."

A few days after the explosion and minutes after firefighters extinguished the blaze aboard the *Mega Borg*, I watched scientists spray oil-eating microbes across a forty-acre patch of floating oil in the first open-sea test of bioremediation. That morning, scientists from Alpha Environmental had mixed one hundred pounds of bacteria with sea water on the deck of a Coast Guard boat. Just after noon, they turned on the pumps, and as I stood beside Dr. Oppenheimer, I watched him turn a hose toward the water, spraying a cream-colored stream of naturally occurring bacteria on the oiled surface of the Gulf.

I remember what a wonderful sight it was, under brilliant blue skies, nearly sixty miles out in the Gulf. I could look in one direction and see the burned hulk of the *Mega Borg*, the flames extinguished. I could look in the other and see an entirely new technology being employed for the first time.

Did it work? Scientists said we'll never know. But from a lay perspective, I can say that two hours later I saw no oil on the water. I saw the same brownish material I had seen in my aquarium and I saw mullets in a feeding frenzy. Several hundred thousand gallons of oil escaped that blazing ship. What happened to it? Did it all burn up in the fire? Evaporate? Is it still on the floor of the Gulf? Or is it possible that the bioremediation worked? All I know is that when we sprayed that stuff on the water, there was oil everywhere. Twenty-four hours later the Coast Guard couldn't find an oil slick, even from a helicopter. And it still hasn't washed up on Galveston's beaches. There hasn't even been an increase in tar balls.

The microbes were an instant hit with the media. *The New Yorker* had a cartoon that showed two microbes with children's faces, one of them in a scowl. The caption read "I don't want to eat my oil." I was even asked to appear on *Good Morning America*. I was eager to show off the advances we were making, and I knew the only way to keep developing this promising technology was to get as much exposure as possible.

A lot of people think oil-eating microbes have to be preserved in a special habitat. The truth is that thousands of them can be carried around in a little plastic bag right in your pocket. I took the late flight to New York and got a few hours of sleep. I was picked up about 4 a.m. and taken to the studio and prepared for my interview. Midway through my conversation with Joan Lunden, I reached into my pocket and pulled out a Ziploc bag of microbes. You should have seen her face.

The *Mega Borg* spill proved one thing beyond dispute: Texas was not prepared to cope with a major oil spill. We could deal with small spills, but we did not have adequate organization or planning to deal with potential disaster. It took two days to get proper equipment on the scene to contend with the fire and start skimming operations. And we were all running around like chickens with our heads cut off. Neither the government agencies involved nor the private sector really established clear lines of authority or responsibility.

And it would probably happen again. More lightering takes place off the Texas coast than that of any other state. In most cases, what the supertankers are doing is putting the oil on tankers that are small enough to navigate the Houston Ship Channel. As our dependence on foreign oil grows, this shipping activity can only increase.

If we needed any more evidence of our dysfunction, it came just four months later. In late July, the Greek tanker *Shinoussa* collided with a three-barge tow at Redfish Island in the Houston Ship Channel. A spill of 700,000 gallons of oil moved into the delicate ecosystem of Galveston Bay— which among other attributes, is the nursery and spawning ground for thirty percent of all the shrimp and other seafood produced by the Texas commercial fishing industry, which is worth $197 million a year.

Unlike the *Mega Borg* spill, which occurred in deep water, this one occurred in a bay barely deeper than a back-yard swimming pool. Cleanup efforts were extremely difficult. Four days after the spill, the oil had moved to the eastern and southern sections of Galveston Bay, coating with crude oil Eagle Point, Pelican Island, and the entire Texas City dike. Smaller patches of oil and sheens were reported moving into the sensitive West Bay. Once again we were bedeviled by delays. We desperately needed a skimmer and two cargo barges from a supplier in Louisiana, but shipment was stalled because wide loads cannot move over Louisiana highways on Sunday. When we finally got five skimmers into the bay, only two could work because of the depth of the water. Based on the facts of this spill, a shallow bay, ringed with sensitive marshes, it became quickly obvious that traditional spill response methods were useless.

The election of Governor Ann Richards and Lieutenant Governor Bob Bullock in 1990 changed the climate in the state capital and enabled oil spill response in Texas to advance from Keystone Cops to state-of-the-art.

The Oil Spill Prevention and Response Act was sponsored by Port Arthur Senator Carl Parker, Corpus Christi Representative Hugo Berlanga, and Galveston Represent-ative Mike Martin. The legislation designated the Land Office as the lead agency in Texas for dealing with oil spills in coastal waters. It gave the state authority to prepare for and prevent spills instead of simply monitoring whatever the federal government chose to do. Governor Richards signed the legislation, and by 8 a.m. the next day, seven oil spills had been reported to my office. We didn't even have a staff.

The oil companies initially were reluctant, but they soon realized Texans expected them to act responsibly and do their part to protect our coastal resources. The oil industry—not individual taxpayers—bore the startup cost of creating a $25 million Coastal Protection Fund. A two cent per barrel fee is assessed on all oil moving through Texas ports. To operate in Texas coastal waters, vessels and reception facilities are now required to show proof of emergency planning and financial responsibility. Following a model in Florida, the state has spill-fighting equipment strategically placed in five coastal districts. The Land Office oil spill team conducts both scheduled and unannounced readiness drills for all industry, state, and local units charged with cleanup responsibilities. Our computer modeling has been upgraded to deal with spills of catastrophic proportions. And the state now has the authority to take civil and criminal action against operators who willfully disregard our procedures of safety and preparedness.

Additionally, the law budgets $1.25 million a year for research by state universities into new technologies. To fully test the microorganisms, we needed to simulate oceanic conditions, and we couldn't do that in two tanks in Port Aransas. In Corpus Christi, Central Power and Light donated six acres so that we could set up a testing facility. To do that, though, first we had to have a fight with federal bureaucrats.

The land is at the end of a peninsula. There's a 750 megawatt gas-fired utility plant, homes on one or two-acretracts, mechanic shops and horse stables. It's hardly a wilderness. There are many cats, some wild and some pampered every night. A biologist with U.S. Fish and Wildlife said somebody had seen a lot of cats on that site and they *might* be ocelots. Ocelots are a small, wild, and endangered species of spotted cat native to Mexico and parts of South Texas. Jaguarundis are another endangered species. On the strength of that rigorous science, the federal wildlife bureaucrats decreed that in order for the Land Office to accept that land, we had to buy and donate three times the acreage of jaguarundi and ocelot habitat. This is why so many private landowners fear the Endangered Species Act. But unlike private landowners, the Land Office has lawyers and biologists in-house who can fight them. It was just a few acres, but it was also ridiculous. We sued U.S. Fish and Wildlife over principle and won the suit. To conduct research into new oil-spill technology, we didn't have to buy a wild cat habitat.

As for the subsequent tests of bioremediation, the crux of the issue was whether a catalyst could be introduced that would make the microbes consume the oil fast enough to make any difference in a marine spill. After two years of study, published scientific papers are now concluding that this naturally occurring process can be accelerated. I plead guilty to believing my own eyes, though I am aware that for some years, a number of scientists have been saying I'm just another politician blowing hot air to get more attention.

Before the *Exxon Valdez* and the *Mega Borg*, the

approach in Texas was basically to wait for oil spilled in the Gulf to hit land. Then day laborers were hired and sent out to clean it up with absorbent paper towels. We've come a long way in just six years.

But the Coast Guard still has its own way of doing things. In March 1996, a barge called the *Buffalo 292* spilled 210,000 gallons of ship's fuel off Galveston. That's not a catastrophic spill by the federal definition; but it did cost $21 million to cleanup. Because of the cost, the responsible party was hesitant to call out a second skimmer that my staff and the then Mayor of Galveston, Barbara Crews, thought was critical.

I flew down to Galveston to meet with the mayor, the local Coast Guard Commander and other local officials to assess the damage and the response. As I flew over the spill site, I noticed that most of the oil was not where most of the equipment was positioned and some of the oil was very close to the beach. After I landed, we drove to the Coast Guard Command Center to meet with local officials. The mayor told me before the meeting that she needed my help in making the Coast Guard and the responsible party more aggressive in the cleanup.

The meeting in the Commander's office was filled with local officials, Coast Guard officers, the media, and the responsible party. I told the Coast Guard Commander that the equipment was in the wrong place and oil was moving toward the beach. The Coast Guard Officer in charge cut me off and said that I was wrong. I told him I had just flown over it and *seen* it with my own eyes, and if he wanted to borrow the state airplane, we'd let him take a look, too. I also informed the Coast Guard that using my authority under state law, I intended to call out additional skimmers immediately to protect our Texas beaches.

Then I asked him when they were going to get containment booms between the oil and the beaches. He said they weren't. They were going to divert the oil to the beaches and clean it up after it washed ashore. All I could think was here we go again. Aggressively, I told him about the tourism industry in Texas, the importance of Spring Break, and that it was not the policy of the State of Texas to use its valuable beaches as booms. Furthermore, I told him that in Texas, it was our policy to react like a firehouse and go big early. I told him he had ten minutes to find every yard of boom in the United States to keep that oil off Texas beaches, or the State of Texas was going to take charge.

He was shocked. I remember the expressions of the young ensigns with him. They were not accustomed to hearing their commanding officer spoken to in that fashion. Though some of that oil reached the backside of the bay, we kept the oil off the beaches. I also remember looking at the Mayor of Galveston and seeing her smiling face.

In Texas, we average about 1,300 marine oil spills a year; 900 of those come from ships or boats. Marine spills move quickly and erratically, and occur in a highly sensitive environment. Our state now has a staff of highly trained professionals who work seven-day work weeks doing announced and unannounced drills, checking contingency plans and equipment, and responding to spills, just like firefighters. Most spills are small in nature and are handled easily; however, the sheer volume of oil and refined products passing through Texas ports is clear evidence that we will have catastrophic oil spills in the Gulf. The only question is whether we're going to be ready to deal with them.

The U.S. Coast Guard is the lead federal agency in responding to major spills in the coastal environment and because federal law guides state law, it still has a lot to say about what happens on our Texas coast. The U.S. Coast Guard is considering weakening critical rules that require those entities most likely to pollute our marine environment to have pre-positioned, dedicated equipment ready for response. I believe that because we have not had a catastrophic spill in the U.S. in some time, the federal agencies are losing their focus and becoming complacent. At the same time, the oil industry feels that preparedness is an unnecessary expense. Adoption of the US. Coast Guard's considered rule changes would mean that Texas once again would not be prepared for a catastrophic spill.

The battle continues.

PART II

CLEAN AIR

8

The Class of '82

In 1982 Democrats put on one of the greatest October stretch runs in Texas political history, and I was pleased to do my part. Going into that election, a few Democrats were popular fixtures in statewide office. At the top of the ballot U.S. Senator Lloyd Bentsen was seeking his third term, and his strength and authority were always a plus. Other Democratic heavyweights were Bill Hobby, the lieutenant governor, and Bob Bullock, the comptroller whose mentorship enabled me to seek office at a relatively young age. But Bill Clements, a Republican Dallas oilman, was supremely confident of winning a second term as governor, and many pundits were predicting a long ascendancy in Texas of the GOP.

Yet when the dust cleared Democrat Mark White was the governor, Bentsen, Hobby, and Bullock won reelection easily, and not one of the statewide office holders was Republican. Jim Mattox was attorney general, Ann Richards was treasurer, Jim Hightower was agriculture commissioner, Jim Nugent was railroad commissioner, and I was land commissioner. The night of our inauguration we all walked together down Congress Avenue. That victory parade was a great time for Texas Democrats. We thought the future was ours.

I had spent my first months at the Land Office learning from Jack Giberson—chief clerk to six land commissioners—and trying to modernize it. We maintained one of the state's most essential archives—a "war" involving vigilantes and wagon train heists was fought in 1842 over whether the Land Office records would be housed in Houston or Austin—yet the millions of maps, land grants, titles, deeds, and surveys had not been inventoried. The Land Office didn't even have an archivist on the payroll, and there was no emergency sprinkler system. A fire in that library would have been a disaster for the state, not just the agency.

One of the most disturbing things was that the veterans loan program was effectively shut down. Why would the Texas Land Commissioner be involved in programs for military veterans? Once more, it's a facet of our state's unique history. In 1836, as the very

first act of the Congress, the Republic of Texas directed Land Commissioner John Borden to issue grants
of land not only to veterans of the war of independence against Mexico—but also Mexico's war of
independence against Spain. Congress's position was
that service in either war was deserving of equal grat-
itude and recognition. So much for interpretations
that the Republic of Texas was anti-Hispanic. From
the very start we were multi-cultural.

So, from that day until we put the rest of the public
domain in the Permanent School Fund, after every war
the land commissioner was responsible for granting
tracts of land to the veterans of wars. We've granted
almost twelve million acres to Texas veterans. During
the Second World War, Texans approved a constitutional amendment that established two things. First,
you cannot issue any debt for Texas without approval
of a constitutional amendment specifically authorizing
it. Second, the legislature can only create a budget that
the comptroller certifies is balanced.

But despite the hard times and occasional financial
chaos inflicted on state government during the
Depression and the Second World War, Texans
thought it was important to recognize and help veterans for their incredible service in that conflict. So in
1945, Land Commissioner Bascom Giles went to the
legislature and the voters with a solution in the form of
a constitutional amendment. It was approved. The
amendment set up a land board that could sell bonds

so that returning veterans of the war could borrow money from the state to buy land. The constitutional amendment authorized the state to incur debt for just that one purpose. The governor and the attorney general served on the board that the land commissioner chairs. That's how the modern Texas veterans program was created.

But by the early 1970s, the program was on a downslide. There was very little money left in the program. Then Land Commissioner Bob Armstrong, who was trying to keep the loan program alive until it could somehow be refunded, had set up a system in which a veteran first had to get a number on a waiting list with 14,000 names on it. But the land board only had enough money to make a couple of hundred loans a year. And in the unlikely event the veteran's number came up, it took our office eighteen months to close a loan.

After I was elected, I'll never forget making a surprise visit to the veterans' division. In order to facilitate the transition, Bob had hired Kay Morris, an efficiency expert and a colleague and friend from when we had worked at the comptroller's office. We arrived at 3:30 on a Friday afternoon, and there wasn't a soul on the entire floor. All the lights were turned off. Kay and I gaped at each other. We were dumbfounded.

That's when I knew how serious the situation was.

When I was a student at Texas A&M, along with other freshman football players, I was encouraged not to join the Corps of Cadets. The coaches' thinking was that athletics and academics were time-consuming enough. And I didn't join the Corps after I stopped playing; I was the first non-Corps yell leader elected in modern times. But it was the late 1960s, the height of the Vietnam War, and I always assumed I was going to be in the service. There were about a dozen friends who ate together at the Sbisa Dining Hall every night. When the Selective Service instituted the draft lottery in 1970, they did it on live television, and just about every young man in the country was glued to a TV set. I said, "I'm not going to watch that. Seeing it's not going to make any difference. Let's go to a movie."

I wound up going by myself to see Clint Eastwood in *Hang 'Em High*. When I got back to my room, there was a tomato smashed on my door. Your birthday determined how high up on the draft list you would be and how quickly you'd be drafted. I soon heard about the shriek that echoed through the dorm when some guy's birthdate was the very first one drawn. My birthday is February 21, and it came out of the lottery bowl number 363. It would have taken World War III for my number to come up.

Precisely because I wasn't a veteran when I got elected land commissioner, I was well-positioned to start rebuilding the program. Nobody could accuse me of doing that to benefit myself and a few army buddies. But I was well aware that a lot of my peers who served in Vietnam were having problems. Among other things, many were finding it almost impossible to buy a piece of land or a first home.

We didn't do anything revolutionary. First we installed an 800-number for Texas veterans. Without identifying myself, I called a couple of times, and it only required dismissing a couple of employees to get the word around that veterans would not be treated rudely. It wasn't that the employees in the division were hostile to the people they were supposed to serve. The problem was that the program was in limbo. One day I was walking out of the building and talking to a longtime employee about my frustration. "Garry," he suggested, "use your head. Folks that work in the Veterans Land Board have jobs as long as we are making loans to veterans. If we run out of money, they run out of a job."

Not only did they not have any incentive to make loans to Texas veterans. In effect they had a disincentive, because they didn't know if there was going to be any funding next year. So we restructured the budget to make the program entirely self-sufficient. And we changed the emphasis in the workplace. We made the program funded one hundred percent by loan repayments and fees paid by the veterans. To revive the program, in 1983 we got the Texas legislature to agree with our plan that no general revenue funds would be used to pay land board salaries. Internally, under the direction of David Gloier, a military veteran who has worked at the Land Office since the Armstrong years, the premium within the veterans' division is on employees who process the applications quickly and efficiently and approve loans to veterans who are good credit risks. As a result, we have become a lending institution that closes ninety-five percent of its loans within ninety days, and our default rate is so low that no private sector lender in the country can match it.

A 1983 constitutional amendment also expanded the program so that veterans could get low-interest notes for mortgage assistance and home improvements, as well as a place in the country. Over the years, we have continued to go back to the voters with constitutional amendments that strengthen and upgrade the program, and they have always supported us. Our philosophy has always been: "We're not doing Texas veterans a favor. We're repaying one."

But I ran for land commissioner knowing that my primary job was to protect the interests of the Permanent School Fund. We had some notable energy battles early on. Depending on the volume of rumors of gold and silver strikes, Texas law had fluctuated back and forth between English precedent, in which private landowners held all the mineral rights, and Spanish tradition, in which all wealth beneath the land was reserved for the king. It really was confusing. I mean, who owned what?

All that accelerated when the extent of the oil and gas reserves in Texas became known. Starting with the first decade of this century, big landowners and big oil companies were always trying to cut the state out of the enterprise, and they tended to get their way in the legislature. In a series of court cases and actions by land commissioners, these efforts were blunted. Trying to get around those court decisions, the big landowners and big oil passed a law in 1931 that came to be called the "Winterman Reservation." The law gave the state a one-sixteenth non-participating royalty on mineral rights of any public land sold—which meant that not only was the state and public education deprived of almost all of any future income, it didn't even have a voice in the negotiations. All business was conducted between the landowners and oil companies. Incredibly, when I took office in 1982 that law was still on the books. With the help of Lieutenant Governor Hobby and House Speaker Gib Lewis, we got the Winterman Reservation repealed.

Early in my first term, I also realized that it made little sense to us that while the Land Office was producing natural gas and selling it at wholesale prices, as consumers we were paying retail prices. In the energy division, I asked what price we were getting for gas produced on state land, and was told it was less than $2 per thousand cubic feet (mcf). Simultaneously, I learned that we were paying $6.75 mcf to heat the building. It didn't take a rocket scientist to figure out that was not a good deal. So we decided to take our gas royalties "in kind," meaning the gas itself, and suggested we could also save other state agencies and school districts money.

Of course that required moving the gas, and the energy companies were incredulous: You're telling us that you want to use our pipelines to sell gas that competes with us? Yes, we said. And what's more, in order to have a pipeline easement across state land, you've got to sign an agreement to transport our gas. Over the years, the in-kind gas program has made additional income for the Permanent School Fund and saved state agencies and school districts more than $100 million.

I believe now that the 1982 Democratic sweep was a vote for optimism, as much as anything. Texas was on an incredible roll, and people were of a mind to look ahead, not back. Mark White successfully portrayed Bill Clements as a figure of the past. But in fact, the industry that Clements symbolized—big oil—was as dominant in Texas as ever. Prior to the early seventies, Texas had effectively controlled the world price of oil. That changed forever when the nations in the Middle East organized a cartel called OPEC (Organization of Petroleum Exporting Countries). OPEC got the rapt attention of import nations in 1973 with an embargo that more than tripled the price per gallon at American gasoline pumps. Ironically, the loss of all that power was at first a windfall for Texas. Following the world energy crisis brought on by OPEC, oil prices took off like a rocket, and the Texas economy was a major beneficiary. The boom brought in droves of people seeking jobs; the population of our state jumped by three million in less than a decade. Profits from the boom laid the foundation of a high-tech industry and set off a spiral of real estate investment. The boom had its own jargon. All at once Texans were talking about "doing deals." "Flipping land." Come to Texas. It's where the action is.

But we had no long-term strategy. For all our talk of diversifying the economy, we were still living in the long shadow of Spindletop and Dad Joiner: The Texas boom was all based on oil production. In 1984, with crude oil selling at an average price of $28.76 a barrel, 14,591 wells were drilled in Texas—an all-time record. You heard giddy and breathless talk of a $70 barrel of oil. Optimism breeds arrogance, and there was plenty of that on display in bumper stickers in Texas. "Oil Field Trash and Proud of It." "Let a Yankee Freeze in the Dark." Then the bottom fell out. Bumper stickers in the Northeast soon gloated about Texans rotting in the sun.

OPEC nations lost the ability to maintain a united front, and with that they lost the ability to control production, supply, and the price of oil. The Saudis put the world awash in oil, and the value of west Texas Crude fell just as fast as it had risen. During one ten-week period in 1986, the price of crude oil swooped

Class of '82: Inauguration night with Agriculture Commissioner Jim Hightower (center) and Treasurer Ann Richards.

Fellow Aggie and friend Henry Cisneros.

Protégé and mentor: With the late U.S. Senator Ralph Yarborough.

Washington lunch with Perry Bass and U.S. Senator Lloyd Bentsen.

Receiving the Chevron Conservation Award from Robert Redford.

Lieutenant Governor Bob Bullock greets young David Mauro.

Fellow activist: The late Congressman Mickey Leland.

President Clinton and the First Lady at the Alamo in 1996 with Ann Richards and (on her left) the Texas Campaign Chairman.

Promoting natural gas cars with Joseph Kennedy and Jake Pickle.

Congressman Kika de la Garza, a staunch coastal ally.

Texas's great open beach defenders, Bob Eckhardt (left) and Babe Schwartz.

Campaign stop with President Clinton and John Sharp.

Dad with David and Alex.

Celebrating MARPOL ratification with Tom Henderson.

Hands Across Texas: Adopt-A-Beach volunteers, 1996.

from $25 to $13 a barrel, and the free-fall didn't stop until the price dipped under $10. What that meant, over the decade, was that 300,000 Texans lost their jobs.

The savings and loan fiasco accelerated, and 250 banks failed in Texas. The real estate market crashed. A new jargon replaced land flips with "see-through buildings"—brand-new office towers with no tenants. Photographs ran in the press of applications stacked halfway to the ceiling on the desks of clerks in the Houston bankruptcy courts. The economy was in a shambles, and Texans who always thought their professions and livelihoods were immune got a rude awakening. Texans were in an extremely dark mood, and who could blame them? Almost overnight our state was struggling through its hardest times since the Great Depression.

And 1986—as my friend Kika de la Garza had observed in the context of beach garbage—was an election year. I remember seeing a very revealing and convincing poll as that campaign shaped up. As someone quipped in a subsequent race, it's the economy, stupid. So much for the celebrated Democratic Class of 1982: White, Mattox, Richards, Hightower—we were all in serious trouble. Better get busy if you want to accomplish more, I told myself. Because you're liable to be a one-term land commissioner.

I felt particularly bad for Mark White. As governor, from our first week in office he had helped me get passed crucial Land Office and veterans legislation. But Mark was not only being blamed for the state of the economy; he was in trouble because he had tried so hard to improve the quality of public education. In shaping the education package, House Bill 72, Mark and his wife Linda Gayle spent a lot of time looking at the educational reform accomplished in Arkansas months earlier by Bill and Hillary Clinton. But Texas and Arkansas are different places. In Arkansas there are only thirty-two counties and two million people. The governor there is almost like a mayor of a mid-sized city. Governor Clinton was able to sell his plan in person.

Also, the Texas Education Agency in Austin took a plan that was laid out clearly by the legislature and made it ten times longer—in the process creating more jobs for more education bureaucrats. Consider what they did with No-Pass No-Play. I remember when I was playing football for Coach John Vasek at Waco Reicher. The coaches checked a grade report every Friday, and if you were flunking a course, you couldn't put on the pads. Not all public school athletic departments in Texas were that diligent, but I know that many had the same system. One week, our punter flunked a test, and we didn't have a backup punter. Coach Vasek held out the punter, and the poor kid sent back there averaged about fifteen yards a shank. The next week, the whole team made sure our punter passed that test. Now the education bureaucrats decreed if the same thing happened anywhere in the

Texas public school system, the student couldn't deal with it the next week. He or she was ineligible for extracurricular activity until after another six-week grade report. That's half the season in any sport. Bureaucrats turned an honest attempt at educational reform into a bureaucratic nightmare for teenaged kids.

I'm proud of the progress we made during my first term. I also made some glaring mistakes. I acted my age. In private life I had joined the crowd of Texans eagerly borrowing money and investing in real estate. First I thought I was smart, and then I thought I was lucky. It turned out I was neither. I held on most of a decade trying to honor those obligations, but I wound up broke and bankrupt, too. After I paid all the debts I could, I wound up with no money in the bank. I was hanging on till payday like a lot of other people in this state. And there's nothing more humbling and reawakening than having to pay your bills with money orders, because nobody will take your check.

The Republicans underplayed the hand they were holding in 1986. I think they were so rattled by the big surprise four years earlier that they decided to put all their resources and energy into regaining the governor's office. Bill Clements won his rematch with Mark White, all right. But Bill Hobby deserves the credit he received for rounding up the wagons and preventing a full-blown massacre. If the Republicans had seen the polls I saw, they would have run hard against every Democrat and sent us all packing. For land commissioner they offered a bar owner named Dudley Anderson, who put his name on the ballot as M.D. Anderson, hoping to be confused with the revered Houston cancer research hospital. Only in Texas.

At the start of my second term, much of my public effort was directed at MARPOL and the other coastal initiatives. But if there was any temptation to forget the primary role of the Land Office in state government, it evaporated the minute we held a lease sale. The energy industry was devastated in Texas, and the wreckage inevitably extended to our educational system. By the end of 1987, the oil bust had cost our public school districts at least $250 million in royalties. The money to fund education had to come from somewhere, and taxation is not a popular option. It was clear that the Land Office—in fact the whole state—was in desperate need of some long-term economic planning.

When energy ministers in the Middle East were designing the OPEC cartel, the model they studied was the Texas Railroad Commission, which regulates energy production in our state. Anyone who spends much time in Texas inherits the great lore of the East Texas, Burkburnett, and Permian Basin oilfields. Gushers and derricks define us as Texans just as much as the Alamo and the Chisholm Trail. But half a century has passed since there was a major oil play in Texas. It's no secret or accident that ever since Alaska's Prudhoe Bay discovery, majors and independents alike have been getting rid of domestic holdings and moving their exploration for oil abroad. The major oil companies now make most of their U.S. money in refining.

When the bust arrived, Railroad Commissioner Kent Hance made some trips to the Middle East, hoping to convince OPEC to bargain with us on equal footing. Kent was dreaming. Sure, there's still undiscovered oil in Texas, but production-wise, on the global register we have steadily been reduced to a bit player. The Texas oil boom of the 1970s and early 1980s was a wobbly construct of manufactured shortage and manipulated prices. Who knows, maybe something like that could happen again. But few Texans who lived through the bust want to get back on that roller coaster.

On the other hand, we have every reason to believe that energy will still be one of our three leading industries—along with tourism and agriculture—well into the twenty-first century. In terms of sheer supply, our energy ace in the hole is not oil, but natural gas. Texas has one-fourth of the nation's proven natural gas reserves. At the Land Office, we knew that if Texas could find ways to sell more natural gas, that would begin to offset loss of oil income to the funds for education, and it could help revive the economy. But that was a large, formidable "if."

Gas had always been the industry's unwanted stepchild. Texans who grew up around those famous oilfields at their height of production remember skies and horizons aglow at night with flares of gas burned off as a waste byproduct. Investors and drillers considered a natural gas discovery better than a dry hole, but not by much. The Federal Power Commission regulated gas prices at the wellhead with erratic results and thin margins of profit for producers. Amazingly, in Texas no well was drilled specifically in search of natural gas until 1967.

Making matters worse, following the OPEC embargo and gasoline shortages of the 1970s, President Jimmy Carter and his energy secretary, James Schlesinger, institutionalized as part of national energy policy a belief that the United States was almost out of natural gas, so it really had no place in that national policy. In fact the policy actually mandated that use of the fuel be discouraged. Carter's energy experts and planners have since been proved just flat wrong, but no matter: from 1973 to 1987, gas usage in this country dropped from 22 trillion cubic feet to 16.7 trillion cubic feet. In the seventies and 1980s, the United States was the only major industrialized nation that reduced its consumption of natural gas.

Waiting in line to reach a filling station pump was not the only new concern in this country. Americans were beginning to find out about acid rain that kills forests and contaminates lakes in Europe, Canada, and the northeastern states. They were learning about global warming—or the greenhouse effect—and ozone depletion. (There are two very different concerns about ozone. If the insulating layer of ozone in the stratosphere gets too thin, harmful rays of the sun are no longer filtered out, greatly increasing the risk of, among other things, human skin cancer. Too much ground-level ozone, on the other hand, contributes to lung disease, asthma, emphysema, and other respiratory problems.)

One of the problems with the air quality debate is that it's couched in abstract and technical language. Like many Americans, I started out thinking air quality was mostly an issue of aesthetics: Polluted air looked and smelled bad. But to put it bluntly, the simple and necessary act of breathing was becoming harmful to our health. According to the American Lung Association, by the late eighties U.S. health care costs related to air pollution had reached $40 billion a year. One hundred ten million Americans lived in areas where levels of ground-level ozone or carbon monoxide violated EPA health standards established by medical doctors.

And over eight million of those at-risk breathers were residents of metropolitan Houston, El Paso, Dallas–Fort Worth, and Beaumont–Port Arthur. With borderline air quality, San Antonio, Austin, Corpus Christi, Victoria, and Marshall-Tyler-Longview were also in danger of slipping into the costly regulatory grip of the federal Clean Air Act. In 1987 Houston's ozone pollution was second only to the incredible smog in Los Angeles. (Still, the municipal and business establishment in Houston maintained a strict policy of denial. There's nothing wrong with Houston air, the argument went. All you've got to do is change the definitions.)

But there was a better solution than just sticking our heads in the sand—an answer that in the bargain would stimulate the Texas economy and generate revenue for the public schools and universities. The principal cause of air pollution on this planet is the burning of coal and petroleum. Natural gas, on the other hand, is the cleanest-burning fossil fuel, and unlike electric-powered cars and wind and solar utility power, we do not have to wait years for gas to fulfill its promise.

I arrived at the Land Office thinking I knew a great deal about natural gas. My guru on the issue was Bob Krueger, whose enduring legacy as a Texas congressman was not poetry recitations but trying to get natural gas and oil deregulated by the federal government. He succeeded in the case of oil. Though Carter's energy policy prohibited construction of natural gas-powered utility plants—nuclear plants were in great favor then—now my staff was telling me about new "select use" technologies that helped clean up emissions by burning a mix of natural gas with dirtier coal and fuel oil. The hotter gas incinerated pollutants before they left the smokestack.

The first contacts I made in the American Gas Association sugggested that in November 1987 I should meet some guys who were coming to a Fort Worth event for House Speaker Jim Wright. They were career Washington types whose uniform was a $2,000 suit, but for the Texas occasion, they were garbed in cowboy hats and boots and bolo ties. At the Fort Worth Club, in the course of making a pitch for gas as a boiler fuel, one mentioned that the gas association had several Broncos running on gas. "Bronco?" I said. "You mean a *car*? A car can run on natural gas?"

When I got back to Austin, I contacted railroad commissioners Kent Hance and Jim Nugent, and within weeks, we had a pilot project organized in which the Southern Union, Lone Star, and Entex gas utility companies converted the engines of eleven state vehicles to run on compressed natural gas.

At the Land Office it still seemed to us the biggest target for making market inroads was "stationary source" polluters (language, in the air quality debate, for industrial and utility power plants.) But in many countries, vehicles—especially fleet and mass transit vehicles—were also running safely, as well as cleaner and cheaper, on engines fueled by natural gas. Much of the groundbreaking work with vehicles was being done in Canada, and the original motivation there was to create new markets for an abundant Canadian natural resource, not to alleviate air pollution. The city of Vancouver, British Columbia, had everything from taxicabs to newspaper delivery vans running on compressed natural gas (CNG, in the energy trade), and Toronto had spent $25 million replacing its old electric-powered buses with a fleet fueled by CNG.

Other nations were motivated by gasoline prices

that are much higher than what Americans have to pay at the pump. Nearly ten percent of the vehicles in New Zealand were then fueled by natural gas. Italy had 280,000 vehicles on the road that could run on either gasoline or natural gas. Russia was another leader.

And we learned that in addition to fuel cost and maintenance savings, an automotive engine fueled by CNG, instead of gasoline or diesel, cut carbon monoxide and ozone-forming emissions in half and virtually eliminated particulates (common soot.) Plus, fifty percent of the air pollution caused by gasoline and diesel enters the atmosphere through evaporation, not exhaust. An engine fueled by CNG is a closed system; there are no fumes. Overall, the reduction of pollutants amounted to about ninety percent. Though producers of coal and fuel oil were distinctly hostile to the notion, natural gas could help clean up the fumes and smoke of industrial and utility power plants. And, let me stress again, Texas is sitting on one-fourth of the nation's supply of natural gas. So toward the end of 1987, the Land Office set out to make another policy connection that would benefit both the Texas environment and the economy: clean air and natural gas.

Win-win, as I often like to say. But we weren't kidding ourselves that this would be an easy sell. Historically, the energy industry and environmental activists were about as congenial as a rattlesnake and a Texan with a long-handled hoe.

9

Strange Bedfellows

By April 1988, the Land Office had launched a modest natural gas vehicle program in Texas. The eleven state agency vehicles in the pilot project were running cleanly, cheaply, and without incident on compressed natural gas (CNG). But we had a larger ambition: to make natural gas the alternative clean fuel of choice on a national level.

Much of the impetus for cleaning up the air was coming from urban centers in the West. That month, at the invitation of Jon Lear and the Institute for Resource Management—an organization then affiliated with Robert Redford—we had gone out to Phoenix for a conference on the national issue of air quality. For dramatic effect, they held the conference in the

Arizona Club, which is on the twenty-fifth floor of an office tower.

On the Friday that we first met, the desert horizon was an ugly brown smudge. Then the workforce of Phoenix drove home for the weekend. The contrast on Saturday morning was startling, and the organizers told us it was like that every weekend. The cloud of vehicle exhaust had dissipated, and now we could begin to see the nearby mountains. By the end of our conference on Sunday, the view of the mountains' beauty was crystal clear.

But the real kickoff of our effort was just a few days away in Washington, D.C., and at the Land Office we were in a working state verging on panic. After months of courtship—often resembling a ritualistic mating dance—we had gotten the American Gas Association and the Sierra Club to agree to help us put on a symposium called "Natural Gas and Clean Air: An Alliance for America's Future." George Lawrence, president of the AGA, and Michael McCloskey, chairman of the Sierra Club, had agreed to appear together and endorse the Land Office's broad premise. That alliance alone was groundbreaking; it was the policy of the Sierra Club not to cosponsor anything. UPI had announced our press conference with a nice story tagged on the wire: "Unlikely alliance to work for natural gas use."

"Strange bedfellows" was how we were described in other news stories.

But the Capital was obsessed with the Iran-Contra scandal and joint hearings in the House and Senate.

Every time you turned on the TV, you saw the tweed cap of John Poindexter, the gapped teeth of Ollie North, and the miniskirts of Fawn Hall. On the issue of MARPOL and beach garbage, we had already learned it was hard to compete with international gun-running, the Ayatollah, and the jungles of Nicaragua. To make any dent in that coverage—and register more than a blip on the radar screen of policymakers in Congress and the Administration—our conference had to have more going for it than novelty and good ideas.

Senator George Mitchell of Maine was not yet the majority leader, but he was carrying the most prominent bill to reauthorize the Clean Air Act, and he had agreed to make an introductory speech at the conference. The Land Office's Mark Glick, a bright young Pennsylvania transplant, Carnegie-Mellon graduate, and former Beaumont refinery worker, had put together an exciting agenda. With session topics ranging from acid rain to curbing imported oil, we had two days of talks scheduled by experts in the field. What we didn't have was the one Texas newsmaker who could bring real balance to the lineup and make the Washington press corps say, huh?

Bob Strauss had just fallen through. Strauss, a long-time adviser to Democratic and Republican presidents, initially told us that he'd be glad to represent the outlook and interests of the Texas energy industry, but then President Reagan appointed him to a bipartisan committee to start trying to solve the problem of the runaway federal deficit. That created a scheduling conflict.

With the MARPOL treaty ratified, I had asked Tom

Henderson to get up to speed on alternative fuels, and now I asked him to help me enlist an industry leader who would attract press attention. Tom does some of his best thinking when he can't sleep at night. He burst in my office one morning and blurted, "T. Boone Pickens!"

"Right," I said. "Boone Pickens." Boone, who then divided his time between residences in Amarillo and Dallas, was known to the nation as a corporate raider—the maverick Texan who almost pulled off a hostile takeover of Gulf Oil. But Mesa, Inc., whose board Boone chaired, had started in natural gas fields in Oklahoma and Texas. If the natural gas industry had a true champion and an American tycoon, this was the guy.

It was a wild day at the Land Office. I called Boone. At first he said he'd do it. Then he called back and said, "Well, I've got a scheduling problem. If I do this, I'm going to have to fly to Washington, back to Texas, and then back to Washington in two days." In other words, it was pretty inconvenient.

My staff and friends assure me I did this, and recalling their expressions of walleyed shock, I suppose I did. "Boone," I told him. "This is real important to me. You're going to have to figure out how important it is to you. Let me know."

Then I hung up.

In Washington, the third week of April is a cherry blossom and tulip show. To us, the malls and grounds of the federal office buildings looked like a Disney movie. Noting our Texas accents, a cabbie told us one day, "town didn't look like this until Lady Bird Johnson took hold." The week of our symposium brought dazzling spring sun one day, snow flurries the next. Boone Pickens was a terrific and forceful addition to the conference. He said his interest was in creating demand for a resource that at the moment was priced so low it didn't make much sense to bring it out of the ground. He hoped to make natural gas a premium product, and it seemed that its potential as a clean air fuel could well hold the key.

In his understated and thoughtful way, Senator Mitchell was just as impressive, and he encouraged us. The late Mike Synar, a charismatic Oklahoma congressman who became one of our closest allies, stressed that the Clean Air Act *would* be reauthorized—a confident prediction that few people in Washington were then willing to make. At our conference, presentations on gas supply and deliverability— and especially the linkage with clean air issues—got the attention of the nation's key policymakers. For the first time since President Carter's strategists wrote natural gas off as a "speciality fuel" and mandated construction of nuclear and coal-powered utility plants, natural gas was back on the table of national energy policy. After that conference was over, you almost never heard anybody talk about the nation being almost out of natural gas. From then on, every credible clean air strategy advanced by every group included a natural gas component.

While we were in the capital that week, I met with members of the Texas Congressional Delegation—

Democrats Ron Coleman, Martin Frost, and Mickey Leland, and Republicans Jack Fields and Joe Barton. Speaker Jim Wright was always very cordial and accommodating when we came to town. I remember his reaction the first time we told him about vehicles fueled all around the world with compressed natural gas. His great eyebrows shot up, and he said, "You mean you can run a car on that stuff I use in my cookstove?"

It seemed to us that making natural gas the alternative clean fuel of choice was an objective most effectively pursued in a national venue. But we soon found out that pushing ratification of an ocean-dumping treaty had been a breeze compared to influencing national energy policy and pushing for reauthorization of the Clean Air Act. While Senator Mitchell carried the clean air bill in the Senate, powerful members and constituencies were at loggerheads in the House.

Southern California has the worst air quality in the nation, most of it due to car exhausts, and an area congressman, Democrat Henry Waxman, was pushing clean air legislation that would oblige Detroit's automakers to start making clean-fuel cars and trucks. The Waxman bill had no champions in the U.S. oil industry, which was making most of its money refining imported crude oil into gasoline and diesel. Nor did the prospect of diminished gasoline and diesel use excite refinery workers, who have a very strong labor union. Equally unenthusiastic was Waxman's fellow Democrat, Congressman John Dingell of Detroit, who carried a version of the clean air bill that protected the interests of the big automakers and the United Auto Workers. And Dingell, whose father had been a force in the House of Representatives, was chairman of the powerful Energy and Commerce Committee.

And that was just mobile sources, as the air quality pros refer to cars, trucks, and buses. President Reagan had recently promised Canada that U.S. taxpayers would spend $11 billion to start cleaning up acid rain, which is caused primarily by emissions of sulfur and nitrogen oxides from large coal- and oil-fired plants in the Midwest and Northeast. Coal producers and union mineworkers were just as hostile to any change of the status quo.

On one trip to Washington, Tom Henderson and I got an audience with the chief lobbyist for the coal producers. A retired military officer, he listened briefly to our pitch for co-firing and other mixed-fuel technologies that would clean up smokestack emissions more effectively than coal scrubbers, then he shot to his feet and chewed me up one side and down the other.

"I don't agree with one thing you're saying," he told me. "I've got to answer to a board of directors. You go back home and report to some governor—in fact, I'm going to call your boss and have your job."

The man assumed I was a bureaucrat; land commissioners are appointed in most states, if they have one at all. "What boss is that?" I asked him.

"Bill Clements. I know him well," he answered sharply.

"Well, sir, the first thing you need to understand is

that the people of Texas are my boss. I'm an elected official. And the last time around, Bill Clements got fifty-two percent of the vote. I got sixty-one percent."

The belligerent lobbyist went crawfishing all over the room.

But by mid-summer, we could see there would be no federal energy policy breakthrough in 1988. There are only thirteen oil-producing states; the other thirty-seven were enjoying the binge on cheap imported oil. Even Aggies can count.

Reauthorization of the Clean Air Act was no priority of a Reagan administration on its way out of office, and congressional Democrats were frozen by conflicting blocs of support, as they so often are. Democrats have never really figured out how to bring environmentalists and organized labor into agreement and make them comfortable with each other. On this issue, the common ground should have been obvious: human health. But nothing as large, difficult, and complicated as renewing the Clean Air Act was going to happen until after the presidential election.

So at the Land Office we shifted the focus of our effort from Congress to the Texas Legislature. Because of the shattered economy, increased natural gas use sounded good to Texans—almost anything hopeful did. During that time, Railroad Commissioner John Sharp and Congressman Joseph Kennedy of Massachusetts sponsored several meetings between Texas gas producers, industry representatives, and New England consumers. Their emphasis was on promoting natural gas as a clean, affordable industrial and home-heating alternative to coal and fuel oil. But while the climate for unprecedented change in our state would never be better, an unspoken rule posed a formidable obstacle against what we had in mind: Never, ever, was environmental protection in Texas supposed to exceed that *required* by federal law.

As our legislative plan took shape, we drew on our long working experience with the oil and gas industry and the favor we had won with environmentalists through the coastal initiatives. My senior deputy commissioner then was a tremendously talented LBJ School of Public Affairs graduate named John Hall. When John had first joined the Land Office staff, I had put him in charge of the energy division. He's the first Texan I'd ever met whose hometown was Washington-on-the-Brazos, where Texas revolutionaries signed the declaration of independence from Mexico. (Now if you blink on the road you miss it.) John was on the staff of Senator Lloyd Bentsen when I got him to come to work for the Land Office, and later, when Ann Richards became governor, she would make him chairman of the Texas Water Commission—the first African-American to head a major commission in Texas since Reconstruction.

During the George Bush–Michael Dukakis race that fall, John took a leave to travel with and advise Senator Bentsen in his vice presidential campaign. Taking John's place at the Land Office was Dan Ruiz, a friend of mine since we'd both worked at the comptrollers' office for Bob Bullock. Dan had recently been a senior aide of Jim Hightower at the agriculture commission. Another invaluable member was our research director, Lee Solsbery, who is now the second-ranking official of the International Energy Agency in Paris, where he conducted a global climate study. Lee kept driving home the theme of a "free market mandate."

None of this would ever fly in Texas unless it made economic sense. In the very first drafts of our legislative plan, we stressed that if the programs did not prove cost-effective, participants could simply opt out of them. If they cost you more money, you didn't have to play.

The legislative package we designed would address both stationary and mobile source usage. The key was that we were not trying to legislate technological solutions that did not exist. Power plants co-fired by coal and natural gas were up and running in Pennsylvania and Ohio, and cars all around the world were running on compressed natural gas. This was not Buck Rogers technology.

Texans saw utility smokestacks and refineries and chemical plants sticking up in the sky like lightning rods and naturally assumed that most of the air pollutants were belching from them. But industrial and power plants were actually cleaning up faster than other sources of pollution—because state and federal law required it. Making big industry and power companies shoulder all the load for additional air-quality improvement was the most expensive way to do it— and of course much of that cost would only be passed on to consumers. What's more, it wasn't fair. Half the air pollution in Texas comes from about thirteen million vehicles on our roads.

We favored compressed natural gas as a vehicular fuel because of the additional value to the school funds. But propane, a liquified petroleum gas; methanol, a petrochemical; and ethanol, a grain-based alcohol, are also proven alternative fuels. The oil and chemical industries and the agricultural community had every right to expect the deck would not be stacked in favor of one particular alternative fuel. And if ongoing technological attempts to make cars run on electricity and hydrogen proved realistic and effective, then we had to explore those options, too.

But we parted company with environmentalists in Southern California and their congressional allies who wanted to force Detroit to mass-produce alternative-fueled passenger cars. Even if they could somehow scrabble up the votes for that in Congress, the nation had no ready system of retail fueling stations to serve personal alternative-fueled vehicles. We argued that, like in Canada and Europe, the emphasis should be on fleets of public vehicles.

Most fleet vehicles traveled about the same number of miles each day, and they parked the same place at night. That made it easier to provide refueling stations. And while a private passenger vehicle averages about 10,000 miles a year, an average fleet vehicle puts over 20,000 miles on the odometer, and most fleets ran in urban areas. So fueling them with CNG or other alternative fuels had more than twice the desired effect in terms of reducing air pollution. Also, this approach

was consistent with the philosophy of "making polluters pay." I believe that's fair.

My staff now walked through the office speaking
this strange new language of "NOx and SOx"—being
the shorthand for nitrogen oxide and sulfur oxide
emissions. I had begun to realize that nobody talked
about air quality but government bureaucrats and corporate bureaucrats. To sell this notion in Texas, we
would have to make it comprehensible.

That fall, just about everyone with any writing skill
at the Land Office worked on putting together a thor-
oughly researched and endessly revised sixty-five-page
book that would accompany our package of bills to the
legislature. "Putting Together the Pieces: The Recap-
italization of the Texas Economy" did not prove to be
just another government white paper. For weeks, the
red pens of Tom Henderson and Lee Solsbery parried
and slashed like swords. Somehow, it came out clear
and coherent. George F. Will, the nationally syndicated
columnist and celebrity of intellectual conservatism,
read a copy and sent us a note saying, "The report is
impressive and will be a guideline for other states that
must undergo similar transitions."

In the meantime, we still had to convince ordinary
Texans. Mark Glick came back from San Antonio one
day with a clear bottle full of super-fine black granules.
Captured in a lab at the Southwest Research Institute,
they represented the amount of particulates released
in the air in one hour by one diesel-powered bus.
Anyone who's gotten caught in traffic behind a bus or
truck knows how foul diesel exhaust can be. Diesel

engines are particularly bad about spewing particulates or soot into the air, and as I said before, converting to compressed natural gas all but eliminates particulate emissions. If you shook up Mark's bottle, inside it the black soot hung suspended in the air. The "black bottle" became the best prop we'd had since the exhibit of ship garbage.

When I made a speech and said, "those dirty diesel buses," it was an applause line. Audiences reacted the same way when I said, "Texans don't want to breathe air they can see, smell, and taste." And though leaders of big oil and the municipal and business establishment in Houston were clenching their teeth, everywhere I could I was saying those things every time I got a chance.

The message was getting across. The more we learned, and the more that people responded, the more my perspective began to change. We had gone into this for economic reasons. Texas was in trouble, and one thing the Land Office could do was to stimulate demand for our supplies of natural gas. Both Dr. Bud Weinstein, then an economist at Southern Methodist University, and Comptroller Bob Bullock had conducted studies that concluded that over five years, a one trillion cubic foot increase in American natural gas use would create between 50,000 and 100,000 jobs in Texas.

But now I also knew that the people who suffer the most from air pollution are children, the elderly, and pregnant women. Two decades earlier, the highest incidence of asthma in America was among old people. Now children were the most afflicted. I was now a parent myself. Respected university medical studies showed that children growing up in the basin of Los Angeles suffered a permanent fifteen percent loss of lung capacity by the time they were twelve years old. How could we tolerate letting that happen in Texas?

As the new year and the ensuing legislative session approached, our team was still short of the most important players—Texas environmental activists and the Texas energy industry. For months we had attempted to enlist the support of both groups—separately. Leslie Tugman, who came to our staff from the Texas Independent Producers and Royalty Owners Association (TIPRO), contributed her considerable charm, knowledge, and talent to the effort. A week before Christmas, we reserved the Guadalupe River Ranch outside Boerne and persuaded thirteen state leaders to spend a weekend with us exchanging background data, concerns, and views. The limestone lodges of the retreat are arranged in a secluded canyon of the clear-running Guadalupe; it's a great place to burn a fire, take a walk, and get away. But we didn't know what was going to happen. We had real fears the thing would blow up in our faces.

In the energy camp were David Biegler, president and CEO of Lone Star Gas; Sy Orlofsky, president and chairman of Intercon Gas; Bill Allison, president of Houston Pipeline; John Watson, senior vice president of Mitchell Energy; Jack Dutton, executive vice president of Tenneco; and John Spence, regional natural gas marketing manager of Amoco.

Environmentalists and public health advocates strolling out under the cypresses were Dr. Dede Armentrout, vice president of the Audubon Society; Dr. George Crawford, an SMU physics professor representing the Texas Environmental Coalition; Dr. Ken Kramer, state capital representative of the Lone Star Sierra Club; Dr. George Smith, air issues chairman of the Houston Sierra Club; Kathy Jacob, air-quality director for the League of Women Voters; Brigid Shea, programs coordinator of Clean Water Action and a future Austin city council member; and Wade Thomason, program administrator of the American Lung Association.

Our goose was cooked if these people wound up yelling at each other. To break the ice, I suggested that we all get together Friday evening and relax—have a cocktail if one was so disposed.

While other Land Office employees fretted that afternoon, Mark Glick was having the time of his life; he drove the car from Austin delivering our special guest. John Henry Faulk was losing his battle with cancer and knew he didn't have long to live, but on the drive to the ranch he regaled the youngster with rip-roaring tales of the McCarthy era and performed pieces of a one-man play he was sharpening and revising.

John Henry is one of the heroic figures of twentieth-century Texas. After the Second World War his humor and satire made him a national star on network radio and the new medium of television—the heir apparent of Will Rogers—until some McCarthyite called him a Communist and successfully got him blacklisted. John Henry eventually won a libel suit that drove a large nail in the coffin of that madness—an event that led to a successful book and movie about the ordeal, called *Fear On Trial*. But he never saw any real money from the jury's award, and his career on a national stage was over. Time had passed him by.

The amazing thing about him was that he never lost his sense of humor and fundamental optimism. No one found more good in America than John Henry Faulk. Unjust ruin of his own life's dreams made him all the more a champion of our nation's institutions—especially the Constitution and the Bill of Rights.

Toward the end of his life, John Henry's favorite routine was a soaring reflection on the drama of the Constitutional Convention. Oh, it was hot that summer in Philadelphia. These were great and earnest men. But still they had failed to protect the basic freedoms of people who had no privilege and power. For all their hard and honest effort, the document they had was flawed...

As we sat around that room at Christmastime in central Texas, John Henry talked and joked and spun that magic web. I watched his audience as he brought the meaning of his stories forward in time. Eyes gleamed and Adam's apples bobbed. He made the connection between James Madison, the First Amendment, clean air, and natural gas seem as natural as the logs flaming in the big fireplace. I've never seen anything like it. By the time he finished, not only was it possible that we could set aside our differences and work for a common good, it was our patriotic duty!

The people who witnessed that performance were charter members of the organization Clean Air Texas. As soon as the legislature convened in 1989, we presented all members with copies of "Putting Together the Pieces," and on the first of February our sponsors introduced two bills endorsed by Clean Air Texas.

Carried by Senator Don Henderson, Republican of Houston, and Representative David Cain, Democrat of Dallas, the first required that starting in 1991, Texas school districts with more than fifty buses, state agencies with more than fifteen vehicles (law enforcement and emergency vehicles excluded), and metropolitan transit authorities had to buy or lease vehicles that were capable of being converted to run on compressed natural gas or other clean alternative fuels. It set a seven-year timetable for eventual ninety percent conversion of the fleets, but waived the requirement if refueling stations could not be established or if the alternative fueling costs exceeded that of gasoline or diesel.

For the second bill we enlisted Bryan's Democratic Senator Kent Caperton, a former Texas A&M student body president and college roommate of mine. The bill carried by Kent and Representative Hugo Berlanga, a Democrat of Corpus Christi, called on the Air Control Board to determine by 1995 if the technology was available and efficient. If it was, then in 1998 large private-sector fleets in urban areas failing to meet national air quality standards would be required to start converting to alternative fuels. In Houston, Dallas–Fort Worth, El Paso, Beaumont–Port Arthur, or any other Texas cities that in the future might fall into non-compliance with the Clean Air Act, the second bill prohibited the use of fuel oil in industrial and utility boilers between April 15 and October 15 of each year—the peak period of ground-level ozone formation—provided those boilers were capable of using natural gas.

For once, public entities, not private businesses, would be the ones pushing technology. Government fleets were going to be the trial horses. I knew my agency was running fleet vehicles on natural gas and saving money. Why shouldn't other state agencies participate, if it saved money and helped clean up the air? I knew that the Garland Independent School District was burning natural gas in its buses and saving money. Why shouldn't the state require other large school districts to do the same? And why not metropolitan transit fleets, since it was being done all around the world?

But at every level, we gave the participating entities an economic out. If changing to alternative fuels was not at least cost-neutral, then they didn't have to do it.

The day we introduced our legislative package, I woke up with a fever and was so hoarse I could hardly talk. But this was our big day. I dragged myself to the Capitol, and as I croaked into a microphone set on a large U-shaped table, it struck me just how broad, remarkable, and strong this coalition was. TIPRO and the Sierra Club. Lone Star Gas and the Audubon Society. The Texas AFL-CIO and the American Lung Association of Texas. Amoco and the League of Women Voters.

The legislation was reasonable, it was market-driven, it was good for the economy, and it would safeguard human health. Both bills passed both houses of the legislature in June of 1989 without a dissenting vote, and Governor Clements signed them. One legislator explained his position on reading the list of names of Clean Air Texas: "If all these people are for this, I'm sure not gonna be the one against it."

10

The Texas Plan

The day Congress convened in January 1990, Speaker Jim Wright asked Tom Henderson and me to come to Washington for a meeting in his office. He had been quoted in the press as saying that reauthorization of the Clean Air Act was the first item on his legislative agenda, but the staffer who called offered no specifics about the reason for the meeting. Whatever Speaker Wright wanted, that's the kind of request you managed to work into your schedule. Our puzzlement deepened when Congressman John Dingell and his top aide, David Finnigan, walked in the room. Following terse greetings, we didn't say one word.

The veteran Detroit congressman had a bad cold. He obviously felt terrible, and neither he nor Finnigan were happy that Tom and I were there. We had met him two years earlier, at the airquality conference held by Robert Redford's environmental organization in Phoenix. We had even gone out to dinner with him. But as the debate on the Clean Air Act heated up and we continued to lobby for American-made cars fueled by something other than gasoline or diesel, we fell completely out of favor with Dingell and his chief of staff. It had reached the point that Finnigan refused to be in the same room with us. Except now he had no choice.

Speaker Wright and his longtime Democratic colleague and rival exchanged the kind of formal niceties that members of Congress have made into a fine art. They asked for details of each other's holidays and listened with great interest; they inquired about wives and children and said to be sure to give all of them their best.

Then Speaker Wright leaned across the desk and got down to business.

"John," he began, "I just gave the nation the Speaker's pledge that we are going to reauthorize the Clean Air Act, and we're going to do it this year. Now, John, I'm going to have a clean air bill. It can come out of your committee. Or it can come out of some other committee. But I am going to have that bill."

It was a very intense exchange, and at one point the Speaker gestured at us and reminded Congressman Dingell who we were. "They've got some really good ideas about natural gas, and gas is important for my state. And if they're right about gas cleaning up the air and creating jobs, then we ought to have it in your bill."

The phone rang in the middle of all that, and it was Speaker Wright's wife calling to talk about having dinner with the Dingells that weekend. So Congressman Dingell got on the phone and called his wife to confirm the date and restaurant. In the middle of this pretty acrimonious exchange, they could turn it off and go back to being friends. Still, Congressman Dingell left the office grumbling. Steam was all but blowing out of Finnigan's ears. Tom and I were elated, of course, and we were awed to have been allowed to be flies on that wall. Without exactly spelling it out, the Speaker let us know that our function was to bear witness—and then go forth and spread that word all over town. But he also wanted to communicate that he was looking out for the interests of Texas.

Jim Wright's speakership was a relatively short one. After he resigned his seat in Congress, and years later when his accuser Newt Gingrich in turn came under heavy fire, I often thought about what we'd witnessed. Speaker Gingrich had plenty of detractors among Republicans, but when he got in trouble, key leaders hung in with him, and he survived. It wasn't that way with Jim Wright's top congressional lieutenants. Democrats such as John Dingell held their silence. I'm not saying it happened because of that one small battle. But I always wondered if it was because he was always insistent that his lieutenants, regardless of the local interest involved, put the national interest first.

That spring, as the debate in Congress proceeded, Land Office staff and I participated in several meetings in Washington and other cities on the role of natural gas and other alternative fuels. Democratic Congress-men John Bryant of Dallas and Mike Andrews of Houston were working with Oklahoma's Mike Synar, California's Henry Waxman, Louisiana's Billy Tauzin, and New Mexico's Bill Richardson—now the U.S. Ambassdor to the U.N.—to make sure that something closely resembling "the Texas Plan" wound up in the federal statute. But big automakers and big oil were working just as hard to stop it.

The leadership of the oil companies tends to come from the refining divisions, because that's where they make the great majority of their profits nowadays. So that's one obstacle in talking about alternative fuels: These are people whose primary concern has always been gasoline market share, who never even think about their natural gas subsidiaries. And the way people get ahead in the major companies is to be there a long time, so they're a little bureaucratic, too. If you start trying to convince them to make fundamental changes, they're going to resist. Look at how long it took them to admit they could make unleaded gaso-line. Until the bitter end, they kept saying they couldn't make gasoline without lead. It wouldn't burn in cars. Of course they could, and it would. When they had to.

When alternative fuels became a hot topic because of the clean air debate, ARCO recognized that almost all of its profits came from selling gasoline in California, which has the worst pollution problems. ARCO had the most to lose from alternative fuels. Suddenly, ARCO unilaterally said it could make a "reformulated gasoline" that would solve the pollution problems associated with gasoline.

All the other companies said ARCO was wrong. Reformulated gasoline could only be produced in small amounts and would cost the consumer fifty cents more a gallon. But when we passed our legislation in Texas, and they perceived this threat from real alternative fuels, all at once the industry was ready and eager to reformulate gasoline as their solution to the problems of our most polluted urban areas. And what's more, it was only going to cost six cents a gallon more than regular unleaded. They must be geniuses, I thought. They'd figured out how to bring down the price forty-four cents a gallon in six months!

The reformulated gasoline measure had strong support in the Senate and especially on the House Energy and Commerce Committee chaired by John Dingell. The approach also had the backing of two influential Texas congressmen—Democrat Ralph Hall of Rockwall and Republican Jack Fields of Houston.

In our push for natural gas, we were making powerful enemies, but we also made some lasting friends. One was David Freer, a savvy Alaskan who worked for Southern California Gas Company and is now regional vice president of its parent corporation. David brought to our cause his considerable knowledge of Congress and inside-the-beltway politics and the clout of the nation's biggest natural gas utility company. Texas, Alaska, California, no shrinking violets here. Though Tom Henderson was the only one who dressed the part, everybody started calling us "the cowboys."

One strong argument for compressed natural gas was that the price was equivalent to a sixty-five cent gallon of gasoline. The Land Office and Southern California Gas commissioned a study by the Radian Corporation to compare the effectiveness of the two approaches then being considered in cleaning up the air. The technical researchers found that in Houston alone, converting fleets to run on natural gas would in a year eliminate enough polluting gases to fill the Astrodome twelve times. My speechwriters loved that image.

Another enthusiastic new ally was David Rivkin, a brilliant and influential assistant to C. Boyden Gray, the general counsel to President George Bush. David's roots are in Estonia—one of the eastern European republics that regained its freedom with the breakup of the Soviet Union. David told us that Boyden Gray had been an advocate of alternative fuels during the Bush vice presidency, while they were working with the competitiveness council which in the change of administrations had become the responsibility of Dan Quayle. Boyden drove a car fueled by methanol, and he was looking for a way to get reinvolved in the issue.

And President Bush had some acquaintance with what we were up to in Texas. During the 1988 campaign Boone Pickens hosted a fundraiser for Bush at his home in Dallas. John Sununu, the campaign manager who became the Administration's chief of staff, fidgeted through Boone's pitch for natural gas and tried to move the candidate on, but George W. Bush insisted that his dad had time to go out to the garage and see Boone's natural gas car and CNG fueling pump.

One day when Tom Henderson and I were in Washington, we got an appointment to meet Boyden Gray. A tall, lanky New Englander, the President's lawyer said he was extremely interested in our emphasis on converting fleets of vehicles. He told us that his family's cable TV business in Connecticut had run its vehicles on propane for years. He told us he was going to do everything he could to get the adminstration behind this and asked us to help them convince key members of the Texas congressional delegation. I pointed out, "Some of the Republicans in our delegation aren't exactly scrambling to get on board."

"Well, you just tell them that's what the President wants," he said.

Tom and I glanced at each other.

"Uh, since I was recently seen campaigning in Texas for Michael Dukakis, I don't think I'm going to have a whole lot of credibility with those guys."

It wasn't clear whether he thought we were Republicans, or that any Texan would be lined up in support of a President from Texas. Whatever the case, Boyden laughed it off and became one of the most important allies we had, but others reacted like somebody had chased skunks into the inner sanctum.

Fascinated, Boyden picked up Mark Glick's black bottle of particulates and gave it a shake—and asked to keep it. We had to ask the lab in San Antonio to make us another one.

The commitment of George Bush to take the lead on behalf of reauthorizing the Clean Air Act was the crucial turning point in its ultimate success. From that point on, it was no longer a question of *whether* a clean air bill would emerge from Congress and become law; it was a question of what form the legislation would take. In June 1990 President Bush addressed a joint session of Congress on the importance of reauthorizing the Clean Air Act. The administration proposal gave big oil and big automakers a lot of what they wanted. But President Bush also endorsed the idea of converting fleets to clean alternative fuels, and he called Texas's alliance of government, industry, and citizens' groups "a model for the nation."

I believe that was one of President Bush's finest hours, and not just because I agree with much of what he said. His political base was Houston. He had cut his political teeth there. But ever since Richard Nixon signed the first Clean Air Act in 1970, the big law firms in Houston had taken responsibility for keeping it from disrupting the status quo. All they had to do was haggle with the definitions and tie up the process in court—and the big oil corporations, the petrochemical industry, the utilities, and the transportation system would be able to cruise or creep right along, doing just what they had always done.

Now the President of the United States was telling the leaders of Houston that they were wrong, terribly wrong. In that speech he identified three urban areas in this country in which air pollution had reached levels of crisis—Los Angeles, New York, and Houston. He said it would take twenty years to restore air quality that did not imperil human health in those cities. President Bush not only promised to sign the Clean Air Act of 1990; he championed the legislation. It was an act of courage and principle. And later, when Dan Quayle was asked to identify the crowning intellectual achievement of their administration, he pointed to the Clean Air Act.

Back home, the Texas Democratic Class of '82 was not exactly thriving. Everybody was in excellent health, but after 1990 all but two of us were back in private life, pursuing other lines of work. In the gubernatorial race, Ann Richards survived a bruising primary fight with Jim Mattox and Mark White, and then she came from behind to beat another Republican oilman, the colorful but self-destructive Clayton Williams. In the agriculture commissioner's race, Jim Hightower was overconfident, refused to do enough fundraising and spend any of what he had on television, and got himself ambushed by a photogenic Aggie and onetime Democrat, Rick Perry. Jim's bulldog insistence on environmental and consumer protection has been sorely missed in Texas state government, though in his subsequent radio career he has continued to have national influence.

Ann Richards would be the first to say that the environment was not her field of expertise or all-consuming cause. The emphases of her administration were education, the criminal justice system, and bringing new industries and employers to Texas. But the first thing she did as a gubernatorial candidate was to make a weeklong boat tour of the Gulf's bays and lagoons and the Intracoastal Canal, acquainting herself with oil spills, beach access, erosion, wetlands loss, runoff pollution, beach debris, and other urgent concerns of Texans who lived on the coast. "There are no downsides with natural gas," she cut off advisers who rambled on about the technicalities of the alternative fuels debate.

Ann's chief of staff was Mary Beth Rogers, whose mentors included Ernesto Cortes, Hispanic Texas' champion of empowering the powerless, and Bob Armstrong, my predecessor as Texas Land Commissioner. But more than anyone, she was influenced by her husband, the late John Rogers. John was a labor leader, but he was far more than that—he made sacrifices all his life in order to do what was right.

In May 1991, one of the most important things Governor Richards and Mary Beth did was name John Hall chairman of the Water Commission. The agency had regulatory authority over water that could be polluted by everything from municipal and industrial effluent discharge to hazardous waste. For decades the Water Commission had been effectively run by the entities it regulated.

Air, water, and toxic waste issues are of course related, and the old system was inefficient and often unfair. In the permit-seeking process, for example, industries and municipalities found themselves wandering in a maze of duplicating and sometimes conflicting regulations. In the 1991 legislative session, Governor Richards and Lieutenant Governor Bullock set out to merge the Water Commission and Air Control Board into a single agency that became the Texas Natural Resource Conservation Commission, under the board chaired by John Hall, my former senior deputy and a charter member of Clean Air Texas. The lieutenant governor asked me to make Tom Henderson available to help Martha McCabe of his staff and Caryn Cosper, who worked for Senator Carl Parker, work on the legislation. The new agency would be responsible for, among other things, implementing and enforcing provisions of the Clean Air Act. For all of us it was an exhilarating time to be involved in state government.

Although the objective of converting fleets of vehicles to alternative fuels was now part of federal and state law, engine conversion shops were not going to make our dream a reality. It wasn't going to happen until Detroit automakers started to make vehicles that came off the assembly line *designed* to run on compressed natural gas or other clean fuels, and we quickly learned that the Big Three have no interest in anything that does not qualify as mass production.

In 1990, General Motors agreed to build 2,500 Sierra pickups equipped to run on either gasoline or compressed natural gas—if we could raise a million dollars from the private sector to pay for crash, emissions, and other certification tests. We did that. Chrysler manufactured the first cars dedicated to run only on natural gas in its series of vans. But the most dramatic opportunity came when GM announced it was going to close one of two plants: either in Willow Run, Michigan, or in Arlington, Texas.

Officials of both cities and states were trying to convince the automotive giant to keep the plants open and thousands of workers off unemployment. Along with Tom Henderson and Mark Glick—both then in the private sector—the staff of the governor's office, the land office, the commerce department, and our friend Roy Spence, we laid out an offer and challenge to GM called "the Texas Commitment." The state was committed to converting its fleets to clean alternative fuels whether the federal government followed through or not. Governor Richards said that if GM would start manufacturing vehicles in Arlington with engines dedicated to run on compressed natural gas, the State of Texas was in a position to insure a substantial market.

Ann also indicated she'd like to talk this over directly with Robert Stempel, the president and CEO of General Motors. The reply came back that such a meeting was impossible; if the head of GM consented to meet with any governor, he would have to meet with all of them. GM sent some executives down to Austin to confer with Cathy Bonner, the director of our commerce department, and me. The meeting was under way when the door opened, Ann walked in, and sat down. She was a very famous person by then, and the automakers reacted with a start. She listened for a while, and then asked to say a few words.

"I have this belief," she said, "that the people who elected me have the rights of customers. If they say they want to communicate with me, it's my responsibility to do just that. Now, the State of Texas is a big customer of General Motors. And it's real interesting to me that the chief executive officer of General Motors says he doesn't have time to talk to his customers. Now, I want you to go back and tell him that the chief executive officer of the state of Texas wants to have a business conversation with him. Would you all do that?"

A few days later, Robert Stempel received a call from the chairman of the Senate Finance Committee. Senator Lloyd Bentsen said there was a friend of his named Ann Richards that the head of General Motors really needed to meet. It would be a personal favor to him. Would it be possible for Mr. Stempel to get on a plane and come to the senator's office?

As things worked out, the agreed-upon Chevrolet Caprices with engines fueled by compressed natural gas never rolled off the assembly line. GM stopped making those sedans altogether. But because of Ann Richards and the Texas Commitment campaign, the Arlington plant is still open and employing Texans today.

11

The Manhattan Project

B y early 1992 I had been involved for several months in the Presidential campaign of Bill Clinton. Yes, I had great respect for my fellow Texan George Bush. But Bill Clinton had been my friend for twenty years. I had known him when he ran for Congress and lost narrowly, the same year I was breaking in as a campaign manager. I celebrated with him when he was elected the youngest attorney general in the country, and then when he was elected the youngest governor in the country. I commiserated with him when the voters booted him out two years later. I encouraged him when he said he was going to learn from that and not give up. When he called me and asked me to help him in Texas, I thought about it and decided: You know, when somebody you consider to be a friend asks for something like this, you can't turn him down and then go on saying you're friends. I had watched him grow, and there was no doubt in my mind that he could measure up to the task of being President. I enthusiastically accepted.

At the start of the race, President Bush was riding high in the polls primarily because of the outcome of the Persian Gulf War. Conventional wisdom was that better-known Democrats were staying out of the race because it couldn't be won; and that a candidate like Bill Clinton was hoping to make a good showing, establish himself as a national figure, and perhaps win the Presidency another time. But those who thought Bill Clinton was running to lose didn't know the man.

When I agreed to head his Texas campaign, I knew he was one of the smartest men I'd ever met. I trusted Hillary's values and pragmatic common sense. And there was never a doubt they had a solid working partnership. I believed that the way Super Tuesday was set up, Texas could put him on the road to the nomination. And we always believed he could carry Texas in the general election.

He knew our state well because he had run the losing 1972 Presidential campaign. And as attorney general and then governor of a neighboring state he understood our people and their problems, particularly those of Texans living in the normally Democratic, voter rich area east of Interstate 45 closest to him. He had returned to our state many times during the twenty years since he had lived here and he had lots of Texas friends.

In order to take early advantage of those close ties

and to demonstrate political strength in a major electoral prize state, we had scheduled a swing through Austin and Houston for late January. We had lined up many Texas political and financial leaders and convinced them that getting on the Clinton train early was the right thing to do. However, the day before his scheduled trip to Texas, the Gennifer Flowers story broke. The national media was in a feeding frenzy and after the disastrous impact on Gary Hart's campaign of a similar story in 1984, many of the political professionals were saying that Bill Clinton's campaign was over. That he would be forced to withdraw in a matter of days.

A show of strength in the face of adversity became even more crucial. We had scheduled a rally at the State Capitol. As Governor Clinton entered the capitol rotunda he was greeted by a large enthusiastic crowd and the Senate gallery was packed to capacity. Bill Clinton has always drawn strength from personal campaigning, from the people who press to shake his hand, and from those who listen intently to his stirring words. Today was no exception. A beaming Bill Clinton forcefully stated his case for why he should be our next President and graciously and gratefully accepted the endorsement of sixty-eight elected officials. From there we went to an Austin hotel where several hundred of Texas top fundraisers and political leaders had gathered for a luncheon, which further raised his spirits. I was elated. How could anyone fail to see that Bill Clinton had that special touch necessary to stir voters to his campaign?

After those events we boarded buses for the trip to Houston. To hear the press and the Washington political pundits talk, you would have thought we had witnessed two entirely different days. They were still convinced Bill Clinton was history. That it was only a matter of time. Listening to that inside-the-beltway wisdom irritated me. I said something my family wouldn't like to see in print and said this guy did not need to drop out of anything and that I knew him well enough to know he had no intention of doing so. Several of the young political crowd standing nearby just rolled their eyes and moved away. Obviously I was some country bumpkin who didn't understand how things worked in the real world of politics.

Of course, Bill Clinton did not drop out of the race, and the television pictures of large enthusiastic crowds in Texas that day certainly didn't hurt. After a second place finish in New Hampshire though, a strong win was important. I had hoped Texas would be the place for that win, but in the Georgia primary one week before Super Tuesday, Bill Clinton won seventy percent of the vote. Our sixty-eight percent finish the next week was just icing on the cake. Bill Clinton was on the road to the Democratic nomination for President.

During that primary season, one small moment made an indelible impression on me. We were in the middle of a Presidential election. The schedule was hectic. The pressure immense. But Bill found time to get on a pay phone and call his daughter Chelsea so he could help her with her homework. Here was a man with his priorities straight. In the end, we didn't carry

Texas for him in the general election. But we pinned George Bush down and made him spend $6 million in his home state. As bad as we wanted to win Texas, George Bush did not deserve to suffer the indignity of being the first President to lose his home state, and he didn't. But we did what we were supposed to do, and Bill Clinton became the first Democrat to win the Presidency without carrying Texas.

Because of our obvious interests in energy issues, a number of Texans helped put together the Clinton energy team. His two top campaign energy advisors were also longtime, hometown friends. Mack McLarty and Bill Clinton had gone to kindergarten together in Hope, Arkansas. They had remained friends, and sometimes competitors, after Bill moved to Little Rock and throughout their early careers. When Bill had gone off to Georgetown and Oxford, Mack had stayed in Arkansas. He later took over the family car dealership and ended up as CEO of ARKLA, the natural gas utility for most of Arkansas and large portions of Louisiana and Texas as well. Bill Burton had run the newspaper in Hope. Looking for a greater challenge, he had moved to Little Rock where he went to work for the Associated Press as a newspaper reporter covering state capital issues. Compared to putting out a small-town newspaper, Bill found the forty-hour week as an AP reporter to be a snap. With a lot of extra time on his hands, he decided to enroll in law school. Later he clerked for Hillary Clinton at her Little Rock law firm and learned a lot about oil and gas. Throughout my tenure as Texas Land Commissioner, I had had numerous discussions with Bill Clinton about energy issues. He asked Bill Burton to give me a call to discuss some of our Land Office programs and to seek my input on putting together the Clinton campaign energy policy. I told him about our natural gas initiatives, about natural gas vehicles, and about how these efforts were good for the energy, economy, and the environment. I gave him a copy of "Putting Together the Pieces," the publication we had written in support of our 1989 Texas Legislative Initiatives. After reading it, Burton called back to say he planned to use it as a blueprint for a proposed Clinton national energy policy and that he had distributed copies to other aides to help bring them up to speed.

He talked with my associates and others to help in developing his plan. Before its release, he called to say the Clinton energy policy would have three primary touchstones: natural gas, renewable energy (like solar and wind), and energy efficiency and conservation. We suggested that the wiser politics might be to reverse the order of those priorities, but we were awfully pleased to see their understanding and acceptance of an initiative we had pushed a long time.

For the last twenty-four hours of the campaign, Bill Clinton had decided to engage in a grueling marathon "victory tour" of the entire country. The Texas part of that tour included a 10:00 p.m. rally in McAllen and a midnight rally in Fort Worth. On the campaign plane between McAllen and Fort Worth, instead of catching a few minutes' sleep, Bill Clinton was moving around the plane discussing policy issues with whomever was awake and would listen. Ann Richards and I were on the plane. He was so hoarse that he could barely speak above a whisper, but sidled up to us and croaked, "I like what you're doing with natural gas vehicles in Texas because it makes economic sense. You can clean up the air and create jobs at the same time. That's what I'm going to do when I'm President."

In April of 1993 President Clinton made good on that promise. He signed an executive order creating a Federal Fleet Conversion Task Force and asked me to chair it. I asked Tom Henderson to be its staff director. Land Office staffers Stroud Kelley, Nancy Molleda, and Daniel Montoya also played key roles. Our goal was to lay out a plan and implementation schedule for using the leverage of federal government fleet purchasing to promote national production and acceptance of alternative fueled vehicles.

In August we issued our first report identifying thirty-eight cities and regions where the effort should be concentrated. But once again, we learned that the federal bureaucracy takes on nothing quickly. The implementation of those recommendations would be painfully slow. However, as it had happened five years earlier in Texas, at least the task force brought key players to the table and got them talking. Members included representative of Detroit's Big Three automakers, the Department of Energy, the Postal Service, the EPA, the General Services Administration, the Transportation Department, the Defense Department, the White House Office of Environmental Policy, state officials from Colorado, New York, California, Louisiana, and New Jersey, utility and energy companies—and Boone Pickens.

One successful result of our recommendations has been the Department of Energy's Clean Cities Program. When we began the task force work, DOE had the beginnings of a program designed to bring together alternative fuel stakeholders at a local level to develop a specific strategic plan for introducing alternative fueled vehicles in their area. Their concept made a lot of sense, but DOE's commitment to that concept was suspect at best. They proudly pointed out that they had one person working on the program—as a volunteer on her own time. The effort was not much more than a press conference and a plaque and we didn't believe that was going to have a huge impact on changing the status quo.

As a result of our task force recommendations, DOE appointed Tommy Foltz, a young Arkansan who had worked in the Clinton campaign and had served as deputy director of our task force, to head the Clean Cities effort. Today, fifty-four communities have received the Clean Cities designation. Most of those communities now have an active, ongoing program to promote the local use of cleaner alternatives to gasoline and diesel.

A lot of creative programs have resulted from that effort. One in which the Land Office has played a central role and which we are particularly proud of is the program in New York City. Working under an innovative memorandum of understanding with the New

York City Department of Transportation, we have helped to place over one hundred natural gas powered taxis in New York's famous taxicab fleet, with plans for at least 500 by year's end. Ford is now making a dedicated natural gas Crown Victoria taxi which will soon be available in other areas of the country as well.

New York City Transportation Commissioner Christopher Lynn and other officials of Mayor Rudolph Giuliani's administration have joined with the Natural Resources Defense Council, New York State Energy Research and Development Authority, Brooklyn Union, Ford, and others to add buses, delivery trucks and even natural gas powered ferry boats to New York's fleet. As it is often said, "If you can make it work in New York, you ought to be able to make it work anywhere."

Here in Texas, our alternative fuels staff under Director Patrice "Pete" Parsons has helped put together strategic plans for using compressed natural gas and other alternative fuels in Dallas–Fort Worth, Houston, El Paso, San Antonio and Austin, as well as developing an alternatives fuels corridor program along IH-10 and IH-35.

Of course, our task force work was just a small part of what transpired in Washington during the first two years of the Clinton Presidency. My friend got off to an erractic start, picked tough and bloody battles, pursued a strategy that resulted in winning votes in Congress by razor-thin margins, bet big on his health care plan, and lost. And as a result he took a beating in the public opinion polls. It seemed at times that nowhere was the President more unpopular than in Texas.

In Texas it wasn't hard to see the electoral rout of 1994 coming. I almost lost the land commissioner's race to a suburban Houston woman whose stated intent was to dismantle every program we had built over the past twelve years. Ann Richards lost her bid for reelection to George W. Bush. Without question, the Texas election was part of what became a national referendum on the first two years of the Clinton Presidency. But that's not the only reason Ann lost. Bush's message was more focused and disciplined, and most important, I think, he possessed the relentless will to win that she had shown four years earlier against Jim Mattox and Clayton Williams.

But there's an interesting element of history here, too. I've been closely involved in Texas politics since 1978. Not once in that time has a gubernatorial race turned out the way the pundits agreed it would. Incumbent Dolph Briscoe wasn't supposed to lose in the primary to Attorney General John Hill, who certainly wasn't supposed to lose the general election to Bill Clements, who wasn't supposed to lose four years later to Mark White, who wasn't supposed to lose the rematch to Clements. Ann Richards wasn't supposed to beat Clayton Williams, and George W. Bush wasn't supposed to beat Ann. If you enjoy politics as a spectator sport, you have to love Texas.

Our new governor now had to make the transformation from campaigning to governing. For him this was a whole new experience. He had helped his dad in political races and I'm sure had offered his views on things he considered important, but now the responsibility of governing lay squarely on his shoulders. He had to deal with real issues in the real world with real consequences for real people. He had to transform campaign rhetoric into public policy.

Looking back after two and a half years of working with Governor Bush, the thing I like most about Governor Bush is that I *like* him. I think what concerns me most is the superficiality of his understanding of many of our state's most serious problems. Because he had never been personally involved in dealing with most state issues before becoming governor, he has little institutional knowledge or conceptual framework for his ideas which, as a result, often lack depth and texture.

Sending appropriate signals to others involved in government is another dimension of leadership. One reason why the lobby never took on our Clean Air Texas Plan when Ann Richards was governor was because she made clear that she would veto any such attempt. However, once Governor Bush refused to heed the advice of fellow Republicans like Boone Pickens and to defend the progress we had made, the big oil, big auto, and big fleet lobbyists were only too happy to push hard to turn back the clock. As a result, our natural gas air quality program has come under continual legislative attack since 1994.

We made some mistakes too. After the rancorous battles surrounding the 1990 Federal Clean Air Act amendments, the opposing sides had been polarized to the extent of bare civility to each other. The oil companies had spent millions of dollars "reformulating" gasoline to make it a less polluting fuel and they wanted it to play a central role in reducing mobile source emissions. Those of us who had been pushing true alternatives to gasoline and diesel, like natural gas, propane, methanol and ethanol, were in no mood to give in. Finding common ground was not anyone's primary objective. As a result the 1995 battle over inclusion of reformulated gasoline and "clean diesel" as acceptable fuels for the Texas Plan fleet program became a needlessly bitter fight on both sides.

Taking on the two largest industries in the world economy—the oil companies and the auto manufacturers—is never going to be an easy fight. Taking them on unnecessarily, without a fallback position, can be sheer folly. While we may have fought the good fight during the 1995 legislative session, we made some hard-core enemies in the process. The result has been a slowing of the progress toward our efforts to clean up transportation pollution—and lingering distrust all around.

I've seen the polls, and I know that average Texans don't particularly care about natural gas cars. But they do want healthy air to breathe and they do want a healthy economy. Political leadership means figuring out how to give them both. It's our job in government to find sensible ways to curb pollution while not overburdening the economy or the daily lives of Texans.

When bureaucrats and the media discuss the implications of the Clean Air Act, they almost always focus on the penalties and dollar costs for failure to comply.

Too often they focus on grandiose schemes which can have enormously disruptive effects on peoples' lives. Demanding car-pooling and employee trip reduction plans for auto-addicted Houstonians is not apt to find a warm reception. But average Texans are also outraged when they find out that ozone pollution basically sunburns your lungs. And that the burning inflicts the most harm on lungs of children and old people. How long would we tolerate dirty air if newspaper headlines blared: Don't let your children go outside because it's not safe to breathe the air. People wouldn't stand for it.

I'm willing to talk and work with anyone interested in trying to figure out the best way to do what needs to be done. But protecting the respiratory health of our children and aging parents requires a clear plan of action. That plan must be fair, must not be overly intrusive in our daily lives, and must be perceived as having a reasonable chance of success.

Devising such a plan is not simple. There is no silver bullet. It must include many approaches, each with small but important benefits. The easy things have already been done. But we have a governor who has no plan at all. His head-in-the-sand attitude toward environmental problems in general and air pollution in particular, and his approach of blaming the federal government for all our problems, may enhance his political career, but it does nothing to solve the problems. Fleet use of cleaner alternatives to gasoline and diesel is a good place to start. Why has Governor Bush failed to embrace an approach that his dad called a model for the nation? You'd have to ask him.

12

Mauro's Mighty Mice

In the fall of 1971, I was halfway through law school at the University of Texas and had just been offered a job in the attorney general's office. It was a good deal, it seemed to me. I'd be working part-time and I could stay in school. I was all set to take it. Then I got a phone call from Ralph Yarborough, the former U.S. senator who had lost to Lloyd Bentsen the year before. He wanted me to come by his office.

Ralph Yarborough was a hero of mine. He had run for governor three times in the 1950s and nearly won, before being elected to the U.S. Senate in 1957, and had always been one of the few Texas politicians who stood up for the working people of Texas. He fought for civil rights when that didn't make you very popular in the South, including Texas. Senator Yarborough lost his seat in a bitter 1970 race to Lloyd Bentsen; now he was trying to decide whether to try to take the seat of Senator John Tower or make one more run for governor. I figured the Senator wanted me to write a speech for him; I had written one a few months earlier that had gone over well when he delivered it to a group of college students.

But it wasn't a speech he wanted. "I want you to go to work for me," he said. "I'll pay you $500 a month. I've decided I'm not going to hire many of the old guys who used to work on my campaigns. I'm going to hire all you new young guys."

I'll never forget it. My first reaction was to say: "Oh, Senator, I'm in law school; I can't." But then I stopped. At that point in my life, I never expected to be involved in politics. Politicians were important people, connected people. I was none of those things. I also needed a job badly, and had one waiting for me that would allow me to stay in school and have more money to live on than Senator Yarborough was going to pay. I said yes anyway. I knew, intuitively, that I had just been offered an opportunity I could not turn down. I remember walking out of the Senator's office wondering how in the world I was going to explain this to my dad.

One of my many chores for the Senator was driving him wherever he had to go. It was in that car that I began to get my political education. He had opted to run for the Senate. Even though we lost that race in a Democratic runoff to Barefoot Sanders—now a Dallas federal judge—and that was the end of my hero's political career, I learned more of value in those months than I ever could have learned in a law school classroom.

On several occasions the Senator told me how, as a young assistant attorney general in the 1930, the big oil companies operating in Texas had applied extreme heat on him and his boss, Attorney General James Allred. Big oil didn't want to have to share its wealth with the Public School Fund and the University Fund, but Yarborough argued the case that saved the funds for the school children of Texas. If not for Ralph Yarborough and his boss Jimmy Allred, who would go on to be governor, the chances of preserving the integrity of the School Fund and the University Fund would have been pretty slim.

The Senator told me that, with the exception of being elected to the U.S. Senate, that court victory over big oil—*Magnolia Petroleum Company v. Walker*, it was called—was his proudest moment in public life. For me, the story was both a warning and an invaluable lesson—and it would come in handy when I took office myself a decade later.

Without Yarborough's crucial legal victory, the Permanent School Fund would not be worth $14 billion today. Nor would the interest provide over $200 per Texas student every year.

Spanish law tradition assumed that all mineral wealth belonged to the king. That meant in Texas, land grants by the crown conveyed only an interest in the surface. English common law assumed conveyance of land in fee meant conveyance of both surface and minerals. In the Constitution of 1840, Texas decided to adopt English common law tradition in most areas, but specifically to retain ownership of all minerals in the Republic.

Periodically, subsequent actions of the Congress of the Republic and then the state legislature, operating on the presumption of public mineral ownership, acted to relinquish or give up that ownership to previous buyers of surface acreage. But the presumption of public ownership of mineral wealth continued to be honored—until a startling reversal of that precedent in 1901 by the Texas Supreme Court. It ruled that any land not specifically designated on the deed as "mineral classified" was deemed to have conveyed both surface *and* minerals.

Land Commissioner Charles Rogan, tipped off the day before the court ruling, took bold action to preserve the state's ownership of the minerals in all of the remaining public domain. He rounded up every clerk in the Land Office and had them come in before work that morning and write "Mineral Classified" in hand on every remaining deed. That simple audacious act saved the state's claim to mineral rights on nearly seven million acres.

In my opinion, Charles Rogan is a Texas hero. A native of Texas, and a member of the first graduating class of the just-opened Agricultural and Mechanical College of Texas, Rogan is hardly known to the general public, but he should be. He's the big reason why we have significant sums of money in our Permanent School Fund today.

A land commissioner should always be judged on

how aggressively he or she is willing to protect the integrity of the School Fund and the public domain. The "Turner case," which became a law suit that could easily have ended my political career, was a shocking assault on that integrity. It presented just such a challenge for me.

In 1982, I won my first term as land commissioner by defeating a popular state senator from Midland named Pete Snelson. Soon after I was sworn in, I went to my first university lease sale, which happened to be in Midland. Midland people were curious about "this kid" who had just become land commissioner, and after the lease sale, at a party sponsored by the University of Texas, some young attorneys asked me to visit with them in the bar.

After a while, they'd had a little too much to drink, and they got to talking about a law firm in Fort Worth that had figured out how to get around the Relinquishment Act, the 1919 law that requires surface owners to act as the state's agent to lease the minerals retained by Commissioner Rogan. In return, according to the law, surface owners got fifty percent of the bonuses, royalties, and rentals from minerals produced for serving as the leasing agent.

The Midland attorneys all winked at each other and laughed, but in essence what they were telling me was that somebody had figured out a way to get around that requirement, so that the surface owner would get more than the fifty percent. In other words, more than their rightful and legal share. The attorneys said if you ever had problems with the Relinquishment Act, everybody knew to call these fellows in Fort Worth.

The very next week, I started making myself a Relinquishment Act expert. I set up a Relinquishment Act Task Force to examine many of the state's leases, and pretty soon we discovered that indeed the state was not getting its fair share of royalties from oil and gas leases. In contested matters, most energy companies and individuals tried to work with the Land Office to balance their accounts. But some times we had to go to court to make sure the interests of the School Fund were protected. And that's how the Turner case came to be.

The story of the Turner case begins in the barren sand hills of eastern Pecos County, a decade or so before the young Ralph Yarborough was fighting to protect the school funds. Before 1926, those scrubby hills of Pecos County made up just about the loneliest piece of real estate in the United States. The land was so useless that surveyors rarely bothered to even survey it in person. They'd just plat their maps from rudimentary field notes.

A fellow named Ira Yates lived in Pecos County. Yates barely scratched out a living as a rancher and grocer, but he and his wife Ann had dreams, particularly after the discovery of oil in 1923 in nearby Reagan County. Yates started hunting for an oil company willing to drill on his land.

He found one, and on October 28, 1926, the Yates No. 1-A came in. Within twenty-four hours, oil company representatives were swarming all over Yates' ranch, and the rancher sold leases until well after midnight. He made $180,000 that first day.

A tent city bloomed—Yates named it Iraan—and rumors spread of boundary markers moving mysteriously overnight, a common occurrence in the first flush of an oil boom. Land changed hands at the toss of a coin.

The field's reserves have been estimated at four billion barrels, and it's expected to keep producing throughout most of the 21st century. In just twenty months, the Yates Field supported more than 200 producing wells; they were so productive, they sent oil prices plunging. For the first time ever, the state stepped in to restrict production.

The Yates Field drew not only major oil companies but also a type of wildcat oilman known as vacancy hunters. Vacancies were lands not covered by existing surveys—gaps between parcels of land inadvertently left by surveyors who weren't all that interested in hanging around the badlands of Pecos County and risking encounters with Comanche and rattlesnakes. Vacancy land, ranging from a few feet to a few miles, belonged to the state and could be purchased by anyone who proved in court that it was, in fact, not legally part of a properly existing survey.

In 1927, a Midland oilman named Fred Turner Jr. and a wealthy Austinite named Bob Reid filed claims on opposite ends of a tract in the Yates field. They believed it contained a 101-acre vacancy (unsurveyed, hence unowned land). They agreed that whomever was awarded the vacancy by the Land Office would split the interest with the other.

In 1930, Turner acquired four acres in the Yates Field from another vacancy hunter, and three more years passed before Turner recorded his title. Meanwhile, the previous owner executed an oil lease on the land.

The Land Office was unsatisfied with the $397 bonus payment the state received under that lease, so as a young assistant attorney general, Ralph Yarborough sued all the parties. The court appointed a receiver to solicit other lease proposals. Turner submitted a lease he had allegedly negotiated with Albert Fasken, a Midland banker. The lease stated that Fasken had agreed to pay a bonus of $20,000, significantly more than the original bonus.

The court named Turner the surface owner and, with a few modifications, eventually ratified the Fasken lease. In addition to the bonus, Turner and the state each were to split a one-eighth royalty interest.

The Fred Turner No. 1 was drilled, and it made Turner an immensely wealthy man. Fasken simultaneously transferred his interest in the well to Turner and to a corporation in which Turner and Reid, along with their lawyers, were the major stockholders. The result was that the Turner group and their heirs received more than $160 million from the well, while the Permanent School Fund got $10 million—hardly a fifty-fifty split, or a fair deal for the school children of Texas.

Turner did all right. He built a huge house on twelve

contiguous lots in Midland and indulged his penchant for thoroughbred race horses. His Tomy Le won the Kentucky Derby in 1959, and his daughter, Dorothy Turner Scharbauer, won the Derby and the Preakness in 1987 with her horse Alysheba. The acreage ended up making Turner millions and his heirs over the years continued to collect on the leases.

In 1967, Turner's former bookkeeper, Andrew Knickerbocker, began writing Texas officials, offering them evidence of a monumental swindle. Lawyers in the attorney general's office reviewed Knickerbocker's claims but could find "no basis under the law" for filing suit against Turner's heirs.

What our Relinquishment Act Task Force found years later was a strawman scheme designed to make certain that the Turner group got more than the fifty percent they were entitled to under the law and the state got shortchanged. The ratio in favor of these supposedly equal partners was sixteen to one.

It worked like this: Oil speculator A would buy a vacancy that he thought had oil on it and where the state owned the minerals. Now, under the Relinquishment Act, what he is supposed to do is get the best price for leasing that land for the Permanent School Fund. That is his fiduciary relationship; that is his duty as a trustee for the school children of Texas. But the prospect of big money, I suppose, makes some people look for ways around their legal obligations.

What some of these people would do was create a strawman to lease the land for a token amount but who actually represented the land owner. So the state would get half of the token minimum lease, and the land owner would end up with the remainder. He would drill the well himself and get not only half of the bonus, royalties, and rentals but *all* the profits from the lease, which in Turner's case turned out to be $160 million. The Permanent School Fund got only $10 million.

Slowly our staff began to piece together a paper trail of this half-century old transaction—examining documents in the attic of the Pecos County Courthouse, exploring warehouses crammed with boxes of files and extracting yellowing deeds, dog-eared production ledgers, and tattered receipts for rental payments. Then they brought in the accountants to try to make sense of all they discovered. Stroud Kelley, our special counsel, called it "forensic accounting." And, sure enough, we discovered the huge gap between what the Turner group made off the land and what the state had received.

We also secured testimony from Andrew Knickerbocker, by then ninety-five years old. Knickerbocker indicated that Fasken had acted as a "strawman" in the deal. Unknown to the state at the time, Fasken's lease of the tract and simultaneous transfer of the lease to an oil company controlled by the Turner group was a total sham.

"Fasken didn't ever receive a dime," Knickerbocker testified in a deposition. "His name was just used as a strawman in the deal, and he had no interest in it."

According to Knickerbocker, something like this happened: Fred Turner walked into Albert Fasken's office, gave him $10,000 in cash, and said, 'I want you to use this money to lease this land from me.'"

It was out of a movie. It was like reading *Giant*.

The arrangement was not recorded in county deed records, because that would have put the state on notice that Turner's arrangement with Fasken had not actually transferred a real interest to him. So Turner not only received half of the one-eighth royalty he deserved as the trustee and agent of the state, but he and his group received huge amounts of money through their ownership of an oil company that operated the two wells on the lease. On March 15, 1988, I urged the attorney general to file suit.

"I was his right-hand man and I did a lot of the paperwork to trick the state," Knickerbocker later told a newspaper reporter. "My testimony made the case for them. I happened to have some pretty valuable documents, which I furnished. I am the only one, I guess, who is still alive and was mixed up in the whole thing."

The old man was right. He had the documents— primarily correspondence between Turner and his lawyers in the prominent Fort Worth firm of Cantey & Hanger. Knickerbocker indicated he was still angry at Turner. "None of that money ever trickled down to me," he said.

I was deposed in a Travis County courtroom by Cantey & Hangar's courtly, old-school attorney, Cecil E. Munn.

"You're a relatively young man, Mr. Mauro, and I think it's obvious that you do have strong political interests and aspirations, do you not?" he asked at one point.

"I have strong political interests, that's for certain," I said.

"Are you interested in running for governor?"

"Some day I'd like to be governor," I said. "I've never made that a secret, but I thought everybody that was born on the banks of the Brazos would like to sit in the governor's mansion some day."

"I would not," Mr. Munn said. "I would hate all those problems, but we all like different things in life. What is the difference between a statesman and a politician?"

"A statesman," I said, "is one who takes on tough issues and acts as a problem solver regardless of the political consequences. A politician is one who only does what is politically positive in his behalf."

Mr. Munn urged me to rise to the character of a statesman. In a prehearing brief, he was a bit more florid. "That's our commissioner—His Majesty Mauro himself," he wrote, "no doubt in his eyes the sophisticate, the knowledgeable, the informed, the clairvoyant—and in clearer eyes, certainly the arrogant Mighty Mouse!"

Mr. Munn's prose nearly took my breath away. Some of my Land Office colleagues made up T-shirts proudly proclaiming themselves "Mauro's Mighty Mice."

Mr. Munn was on a tear. "Behold!" he wrote. "There charges onto the state a new, young land commissioner with unabashed high political ambitions. He is breathing heavily. He becomes quickly judgmental, decreeing that citizens held in highest esteem in their lifetimes by their peers were nothing but scoundrels bent upon cheating the State. To inject an added shot of emotion into the caldron that will contain his witches brew, he phrases his charges in terms of 'cheating the SCHOOL CHILDREN of Texas.' Who could abide a man that would pick on little children!"

I understood what Cecil Munn was doing. He was doing his job as a paid advocate; he was also defending the honor of his firm, old friends, and widowed spouses.

Ignoring his sarcasm, we argued that, under common law, an agent is barred from acquiring an asset he is commissioned to sell for another. As my friend Stroud Kelley explained at the time, "If you ask me to sell a car for you, and I wind up with that car, it is presumed that I have defrauded you. We always felt those schoolchildren were done wrong, and we were going to fix it."

Ralph Yarborough also testified. He told the court that the state knew nothing of the transactions that gave Turner anything beyond the one-half bonus and royalty.

We asked the court to make the school fund whole again. The Turner group heirs argued that a 1990s court could not undo what a 1934 court had determined, and they claimed that it was fundamentally unfair to litigate the validity of a sixty-year-old lease.

State District Judge Pete Lowry accepted their arguments and threw out our case. As you can imagine, it was not a happy day at the Land Office, and our unhappiness continued when an appeals court affirmed his ruling. But then in March 1993, the Supreme Court voted 5-4 to reverse the decision and ordered Judge Lowry to proceed to trial. The majority opinion essentially foreclosed all the defense's legal arguments, and the Turner heirs agreed to settle.

The Supreme Court's majority opinion, written by Justice Raul A. Gonzalez, said in part: "Within the scope of the Relinquishment Act, however, the surface owner is the state's agent to the extent that the state's assets are entrusted to the control of the surface owner, who must not abuse that trust. It is well settled that an agent charged with selling an asset for the owner owes a fiduciary duty to the owner, and that the agent violates this duty by acquiring the asset for his own benefit."

The Supreme Court had said it all. In order for a surface owner to earn fifty percent of the bonuses, royalties and rentals, he or she had to get the best return possible for the School Fund, and if any side benefits were negotiated, fifty percent of those benefits also belonged to the School Fund. Otherwise, according to settled law in Texas, a fraud had occurred.

The State of Texas received $5.3 million up front, plus an increased share of future proceeds. Based on the field's projected oil and gas reserves, we also believe that as much as $250 million more may be recoverable in a dozen or so similar cases.

Although the judgment in the Turner case hurt innocent bystanders—heirs in the Turner group who had nothing to do with defrauding the School Fund—the integrity of the Fund had to be protected. That was my job as land commissioner, regardless of whether we offended a powerful law firm in Fort Worth or a group of wealthy West Texans. I never felt I had any choice.

13

The Big Eight

Bob Armstrong, my predecessor in the Land Office, is one of the most congenial men you'll ever meet, but even Bob had to go to court to make sure the school children of Texas got what they deserved. Bob, along with the State of Louisiana, sued the federal government on grounds it had used the two states to prove the worth of offshore oil production blocks adjoining Texas and Louisiana Gulf waters, and then deprived the states of rightful shares of that production.

The suit was still on appeal when I took office, but a few weeks later, Congress passed a law that settled it. The federal government would pay the State of Texas nearly a billion dollars and a future share of income from the first tier of production blocks adjacent to the state waters.

In the Texas Senate, Lieutenant Governor Bill Hobby celebrated the windfall and asked Treasurer Ann Richards to put the money in the state's General Revenue fund. Attorney General Jim Mattox went along with the idea. I didn't—and neither did my old boss, state Comptroller Bob Bullock, who wouldn't release the funds.

It was clear to me that the revenue from the suit was intended to benefit the School Fund, not the legislature's choice of programs. My fellow Democrats brushed off my arguments, so I felt I had no choice. I hired a private law firm and sued all three of them. And that's when the calls began.

"Don't you know Democrats don't sue Democrats?" I was asked, forcefully. Ann Richards called from the Treasury: "Don't you know Bill Hobby wants us to go with General Revenue?" I told Ann I was aware of that. "Well, don't you know that Bill Hobby is the most powerful Democrat in the state?" Ann persisted. "You can't *not* do what he wants."

I was painfully aware of everything my fellow Democrats were telling me. I knew Bill Hobby was not only powerful; he had been a visionary lieutenant governor. I was just as aware that I wasn't winning any votes among them as the most popular Texas Democrat. Nevertheless, I had a simple answer: That money was owed to the school children of Texas. When the people of Texas elected me as land commissioner, they gave me the job of protecting the Permanent School Fund and its assets.

"A judge may decide you're right," I told Ann and the others. "A judge may rule that this money ought to go into General Revenue, but the fact is, we wouldn't be getting this from the federal government if Bob Armstrong hadn't sued on behalf of the School Fund. We wouldn't have won nearly a billion dollars if he hadn't been protecting the fund. And as land commissioner, it's my job to make sure that money stays with the fund until a judge tells me otherwise."

Well, a judge never told me otherwise, and eventually we settled the suit. Equal thirds went to General Revenue, to the Available School Fund, and to the Permanent School Fund. Two-thirds of all future revenues would go to the Permanent School Fund.

It happened again in 1995. I had to go to the courthouse and let the judges decide. I invited representatives from eight of the biggest oil companies in the world to my office: Mobil Oil, Chevron, Texaco, Marathon, Phillips Petroleum, Exxon, Shell Western and AMOCO. (Later, we added Union Pacific Resources Co. to the proceedings.)

I served rolls, juice, and coffee, then got down to business. "I wanted to call you in to tell you personally," I told those who showed up, "that the General Land Office is suing each of you for royalty underpayments."

They didn't stick around. They set down their coffee and pastries and raced out of the office to call their bosses. Less than half an hour later, our legal staff was in state district court in Austin filing papers that could eventually put millions of dollars into state coffers and the pockets of individual Texans.

Like every politician in Texas, I was well aware that a good way *not* to get elected to higher office in Texas is to antagonize big oil. I knew that by filing a lawsuit against them, I was inviting a pack of trouble from high-paid executives, lobbyists, and lawyers. But I also have a duty to the schoolchildren of Texas and to the taxpayers to make certain Texas was getting the money it had coming.

Although I sued on behalf of the Permanent School Fund, that wasn't the only problem big oil had that morning. The University Board for Lease, which oversees leasing on 2.1 million acres for the Permanent University Fund, voted to join the lawsuit. Late in August, my fellow Aggie, Comptroller John Sharp, also asked to be included as an intervenor. The comptroller's office collects taxes on oil production.

How much all this would mean was hard to quantify, although a 1994 study commissioned by the land offices of Texas, New Mexico, and Colorado concluded that oil companies had been underpaying royalty owners by three to six percent. I knew that between the time I took office in 1983 and 1996, we had deposited more than $2.7 billion into the Permanent School Fund. The revenues from the school lands are invested, and the proceeds are distributed to each school district. If the three percent figure is correct, then that means we had been underpaid at least $30 million. Applying the same formula, Texas would have added $10.8 million to the $362 million it collected in severance taxes in 1995 alone.

How much the state's estimated 50,000 private royalty owners would have lost by being underpaid by the same amount is unknown, but in 1994, Texas produced 538.4 million barrels of crude, which had a value of $8.1 billion. Assuming a one-eighth royalty and the same three percent underpayment, you're looking at an underpayment of at least another $30 million for royalty owners.

Our dispute with the oil companies centered around a disparity between something called "posted" price and the market price for oil. The concept of a posted price came from the days when competing oil companies would literally tack an offer to a property owner's fence post. Until OPEC emerged, posted prices probably represented the real market price of oil. After OPEC, however, an oil futures market came into being, which allowed investors to call New York and get a quote on the world price of oil, just as they could get a quote on pork belly futures or corn futures. It quickly became obvious that the quoted price for oil delivered to the Port of New York and the posted price were rarely the same. The posted price was consistently lower, even when transportation was taken into account. So, which one represented true market value? That was the question.

In 1987, we decided that we would begin to accept part of our royalties "in kind." We would then sell the crude we received by sealed bid at auction. As far as we were concerned, posted prices were meaningless; to get the oil we had for sale, you had to bid for it, and the highest bidder got it. At each auction, we received substantially more for our auctioned oil than posted price—as much as $1.95 per barrel more. As you can imagine, that disparity raised serious questions. I wondered why the auctions were doing so well.

More evidence of underpayment walked in my door. In 1993, an official of Atlantic Richfield Co. showed up unexpectedly and handed over a check for more than $500,000. He said it represented the actual price for which the company had sold its oil royalty, plus interest and a penalty. About the same time ARCO paid us, the company also was paying approximately $20 million to other royalty owners.

I heard rumors that the payments were made after a group of traders who had been laid off in the energy slump of the late 1980s threatened to reveal that the company was selling the oil for more than it paid royalty owners. The ARCO people told the press that the adjustments resulted from a routine internal review. Whatever the reason, ARCO's posted prices began to more closely track the New York price. A few other companies also amended their pricing practices.

At the next Western States Land Commissioners Association meeting, I discussed with my counterparts in New Mexico and Colorado the possibility of hiring a research firm called Summit Management to study one oil field in each state to determine whether the states were receiving market value for their royalties. They agreed that a study would be worthwhile. Summit found that while Texas was being underpaid between three and six percent, New Mexico and Colorado were being underpaid as much as twelve percent.

At about the same time, California and Alaska had decided to go to court because of a much greater variance between posted prices there and New York prices. Alaska was able to show that the moment Alaskan oil was loaded onto an oil company tanker, the company began trading it on the New York Mercantile Exchange (NYMEX). Similarly, California was able to show that as soon as oil began flowing through the only available pipeline, the company began trading it in New York as well.

In Alaska, the oil companies settled their oil royalty litigation for $3.7 billion; in California, they settled for $350 million. (Parts of their suits are still outstanding.)

ARCO's decision to volunteer the information that it had been getting more for its oil than the company had paid in royalties, the findings of our oil-pricing experts that we had been underpaid between three and six percent, and finally the big victories by California and Alaska all pointed to an inescapable conclusion: It was time for Texas to go to the courthouse and let the judges decide what the market price of oil was and whether the Permanent School Fund and the state of Texas had been properly paid.

Accomplished lawyers can be found throughout

state government, but the fact is, the big oil companies' legal talent would easily have us outgunned. Their beginning associates make more than the state's highest-paid attorneys. It was obvious that our decision to take on eight of the biggest oil companies in America meant taking on scores of the highest-paid attorneys in the world. My counterparts in California told me that their oil-company litigation had taken them eighteen years; using the state's own attorneys and paralegals had cost California more than $100 million.

For the State of Texas to win this lawsuit, we had to find the best lawyers—and we had to find a way to fund the suit. Luckily, we were able to convince the Houston firm of Susman & Godfrey, one of the best antitrust firms in the country, to take our case and to fund all expenses. The attorneys even agreed that if we won the case they would allow the judge to decide what their fee should be, to be collected from the losers. If the state lost, we would owe Susman & Godfrey nothing.

Not only were we able to provide the State of Texas and the state's school children with the best lawyers in the country, but we chose to file the suit as a class action. Most Texans realize that there are a lot of wealthy royalty owners in Texas, but they also know that the majority are not. Most inherited their royalty from the family's 180-acre farm. Now, as part of a class action, they too would be able to stand up to big oil.

When I took office in 1983, I spent a lot of time talking about partnerships with Texas oil companies. I told executives in the industry that we wanted them to make hundreds of millions of dollars—so that we could make our twenty-five percent.

But I've learned the hard way that the oil companies are not going to do anything that you don't make them do. They want to be partners, all right, as long as the terms favor them.

"Mauro is just on a publicity campaign," one oil executive told reporters when the Big Eight suit was filed. But nearly a year later, the Texas Supreme Court declined to hear arguments from the Big Eight.

In May 1997, lawyers for Chevron and the Land Office announced to the judge that they had reached a settlement. The settlement establishes a new formula of pricing the oil for which Chevron royalty owners are paid: the NYMEX (world price) of oil, with a *Platt's* adjustment for oil type, minus transportation costs. The Big Eight royalty suit is by no means over. But the price set by the Chevron settlement is significantly higher than what the other oil giants now pay. More settlements are expected. Education in our state—and ordinary Texans with an interest in an oil well or two—may get a fair shake after all.

PART III

LINE IN THE SAND

14

Don't Mess with Coastal Texans

I first ran for land commissioner fifteen years ago knowing that Texans who live along the islands, peninsulas, lagoons, bays, and wetlands had tried and failed to get their communities and natural heritage the protection of a program called Federal Coastal Zone Management. Texas was one of only two states with an ocean coastline that did not participate in the program. The money which was passed down through federal grants was modest—a few million dollars each year—but accepting it did not subject the states to the dreaded entanglement of the Washington bureaucracy. To the contrary, Coastal Management enabled states to influence and even veto projects that could otherwise be rammed ahead by federal bureaucracies such as the U.S. Army Corps of Engineers. But the old guard in Texas was determined not to have it.

Here's how the program began. In the 1960s several hurricanes struck the Gulf and Atlantic coasts—Hurricane Carla was the worst one in Texas. Then, as now, residents of the battered communities expected the federal government to respond to each catastrophe, and the government did, time after time, to the tune of several billion dollars. Though a hurricane wreaks havoc anywhere its wild forces choose, the thinking in Washington was that much of the devastation was preventable; some of the property damage and even deaths and injuries resulted from foolhardy planning. The U.S. Congress didn't want to halt coastal development, but it couldn't afford to keep bailing out states and communities that let people do stupid things—like put up multimillion-dollar developments in storm washover areas. The premise of the legislation was to fund a program that encouraged the states to develop a strategy of responsible development that might keep every storm from becoming a full-blown disaster.

The Federal Coastal Zone Management Act of 1972 was co-authored by Texas U.S. Senator John Tower, a diligent conservative and the first Republican elected to statewide office since Reconstruction. Senator Tower's thinking was: We'll give the states money to "do the right thing" so the federal government and taxpayers won't have to keep on picking up tabs caused by boondoggle and folly. And, once a state develops a plan to "do the right thing" on the coast, we'll even make federal agencies abide by the state's plan. We all know you can't make a federal agency do something contrary to federal law, but within the confines of that law, you can make federal agencies conform their policies to those of the states, cities, and counties. The bill was part of "the new federalism" ballyhooed by President Nixon, and in fact it was a fairly remarkable piece of legislation. This was not a wild-eyed liberal environmental scheme.

But in Texas there were problems from the start. For one thing the name of the statute contained the noun "zone," which raised the specter of the verb "zoning." The legislature in Texas rigorously refuses to give the counties zoning authority, and the coastal community of Houston is well-known as the nation's largest city that has no zoning ordinances. But legislators wanted to pursue the federal program and revenue source, and in 1974 they designated the Land Office to do so.

Bob Armstrong and the Land Office got a $6 million grant from the federal government and tried to establish a system of multi-jurisdictional boards to bring more coherent planning to the coast. There were champions of coastal management, federal or otherwise, in the communities along the Gulf—leaders like State Senator Babe Schwartz of Galveston and Ed Harte, publisher of the *Corpus Christi Caller Times*, Sharron Stewart of Lake Jackson, and Joe Mosely of Corpus Christi. Unfortunately, the Land Office turned over much of the central planning to a group of consultants who operated out of an office on Guadalupe

Street in Austin. Which meant that in the coastal communities, the plan did not inspire a great outpouring of public support.

On top of that, traditional thinkers in the business community, particularly the oil and gas and petrochemical industries, figured that the Federal Coastal Zone Management Act was a threat to them. Exxon, Gulf, Dow, Monsanto, and other corporations owned hundreds of thousands of acres along the Texas coast. A lot of those tracts were floodplains and wetlands. They feared that any new "zoning" authority would declare their property environmentally sensitive and get in the way of doing business as usual. Following the procedure set out by John Tower's bill, state officials drew up a plan, and the governor submitted it to the Office of Coastal Zone Management, a sub-agency of the National Oceanic and Atmospheric Administration in the Commerce Department. Bob Armstrong did his part, and Dolph Briscoe and Bill Clements both signed plans, perhaps to keep the grant money coming. But neither followed through and actually submitted the plan to Washington for federal approval. Coastal Management in Texas was effectively pocket-vetoed by a Democratic and a Republican governor.

When I took office in 1982, I found that I was a member of something called the Texas Energy and Natural Resource Advisory Council. It was described to me as an inter-agency planning mechanism, but after attending a couple of meetings, I could see that it's real agenda was to carry on a cold war against Federal Coastal Zone Management. But 1983 brought Hurricane Alicia, and the state senate responded by conducting hearings about the need for more coherent planning on the coast.

I sat down with my staff and asked them to help me determine exactly what the federal law was supposed to do, what it was intended to protect, and how we could accomplish its goals without having a federal coastal plan. We wanted to facilitate rational development and protect people's rights. If a citizen owned a piece of land and wanted to build a pier out into the water, then he or she should do it in ways that did not restrict navigation or limit their next door neighbors' access. We wanted to start addressing erosion problems, such as Galveston's disappearing beach. We also wanted to get wetlands and other sensitive areas into the state's inventory so they wouldn't be lost to future generations.

Texas is lucky to have an innovative and gifted conservationist named Andy Sansom. He's a loyal Republican, by the way. Andy is now the executive director of the Texas Parks and Wildlife Department and has worked well with governors of both parties. But at that time he was head of the Texas Nature Conservancy, a private group that relies on wealthy donors and pursues conservation goals by making trades and purchases in the real estate market. I met with Andy and mapped every sensitive natural area along the coast; then we decided on independent strategies to get those tracts into the inventory of the state, the Nature Conservancy, or some other private-sector conservation group.

Our final report broke them out into the ten most sensitive areas, the second ten, and so on. By the end of the 1980s almost all of that land was protected in the inventories of Parks and Wildlife or the private groups. President George Bush deserves much of the credit for that.

When he was running for the office in 1988, he said he would institute a federal policy of "no net loss of wetlands" throughout the United States. That set off a lot of grumbling within his own party. But in Texas, energy and petrochemical corporations read the drift of public opinion and decided not to fight the policy. They planned to build refineries and chemical plants and other development. Why dig in their heels and insist on building in migratory fowl rookeries or saline and freshwater marshes? They could exchange the coastal land they owned at a fair market price and easily move their development a few miles inland to acreage that was not a critical part of the Texas natural heritage. So they did.

By then the Adopt-A-Beach cleanups were a success, and the fight over the MARPOL treaty had raised awareness throughout the state about coastal issues. One day Pat Hallisey, the executive director of the Galveston County Parks Board, came to see me in Austin. Pat said it was the first time in his memory that state government was really paying attention to the wishes of coastal communities. He'd been talking to local elected officials, and they hoped the state would seize this opportunity and try to revive participation in Federal Coastal Zone Management.

"Pat," I said, "you know my predecessor drew up a plan to bring Texas into that program. He submitted it twice to the legislature, twice to different governors, and he didn't get our state into Federal Coastal Zone Management. What he got was his Land Office budget line-item-vetoed twice and his budget reduced thirty-five percent over a three-year period. Public land management was set back twenty years in this state because of the federal coastal zone fight. I don't think I want to get back on that horse."

Pat swore that times had changed, and that they were going to pursue a strategy to build a local consensus. They were going to approach their state senator, Chet Brooks, and ask him not to go out on a limb to pass a plan, but to create a study to determine if the climate was right for building consensus and starting over.

Certainly we needed to pay more attention to the coastal region. The coastal population had grown from 600,000 in the early 1920s to 4.5 million by the end of the 1980s. There had been a thirty-five percent loss of coastal marshes since the 1950s—to development, pollution, erosion, and subsidence. The coast al region accounts for sixty-five percent of the nation's petrochemical production today and a quarter of its refining capacity. Nearly $6 billion is spent each year by tourists in the eighteen coastal counties. Since the first offshore lease was granted in 1922, oil and gas platform wells have generated more than $2 billion in oil and gas revenue for the School Fund. The Gulf's commercial fishery, eighty percent of which is shrimp, is worth $197 million a year.

Hurricanes are not the only unpredictable force of nature, and the Texas coast has more than its share of easily swamped structures and dubious planning. In 1954 in one of the most controversial planning decisions made by coastal residents, a hunting and fishing organization in Beaumont bought land on Bolivar Peninsula to cut a new pass between the Gulf and Rollover Bay. They wanted to improve circulation and sportfishing in the bay, a reasonable objective which the new inlet accomplished. Just about anytime now you'll find the anglers lined up shoulder-to-shoulder along the cut, which is called Rollover Pass. But many area residents are convinced that the inlet exacerbated beach erosion on Bolivar Peninsula, a process that has resulted in homes half-submerged and abandoned in the Gulf. Whole blocks of onetime neighborhoods are now under water.

In Matagorda County there's a small community of seaside homes called Sargent Beach. But the strip of land where they're built is no barrier island, and nature did not intend this particular shoreline to be a sand beach. The illusion of an island was created by a nearby cut through the mainland for the Gulf Intracoastal Waterway—part of a highway for small crafts and barges that extends all the way from Brownsville, Texas, to the Chesapeake Bay, on the Atlantic.

At Sargent Beach erosion has worn away the land so

much that you can stand on the deck of a boat in the canal and see the waves in the Gulf breaking just a few hundred yards away. One direct hit by a hurricane would not only sink the homes, it could easily breach the waterway, creating a transportation disaster. Small barges on this narrow coastal channel move over 17 million tons of goods a year in Texas—commerce worth more than $20 billion. No amount of dump-trucked sand or dredging material on the shore would convince the Gulf's tidewash and currents that this man-made island was a natural barrier of beaches and dunes. The only solution, it eventually turned out, was an eight-mile-long $60 million seawall. These are the kinds of problems that city and county governments can't solve without financial help and well-coordinated planning.

So in 1989, legislation was co-authored and passed by Senator Chet Brooks of Galveston and Representative Ralph Wallace of Houston to authorize the development of a comprehensive coastal management plan for the state's coastal public lands. To ensure that the recommendations would reflect the viewpoints and needs of all Texans, we recruited an advisory committee on the coast; anyone who had the time, desire, and energy could serve. We had several weekend work sessions, and we hired some facilitators from Texas A&M. I was still skeptical that this could really get off the ground, but I remember going to one of those retreats in Clear Lake. They had pulled together representatives from the real estate community, the oil and gas industry, the Chemical Council, environmental groups, and local government officials. I began to think it was possible. The first thing we did was stop using the word "zone." It was amazing how much that did to change the tone of the discussion. We kept talking and agreed the process had to be grassroots driven: We had to figure out a way to ensure that the program didn't become the captive of Austin or Washington bureaucrats. In 1991, in a report to the Texas Legislature, the grass-roots advisory committee came up with three broad issues of concern—coastal erosion and dune protection, wetland loss, and beach access. As a result of this consensus, the legislature passed Senate Bill 1053 authored by Senator Chet Brooks of Galveston and Representatives Mike Martin and James Hury, both of Galveston.

Senate Bill 1053 made the land commissioner chairman of a seven-member Coastal Coordination Council that would develop and implement the plan. Members included representatives of the Parks and Wildlife Department, the Railroad Commission, the Natural Resource Conservation Commission, and the Attorney General's office. The governor appointed two members who had to be coastal residents—a citizen and an elected city or county official. The legislature said to move forward with the development of a Texas Coastal Management Plan, but do it with no additional money and no new staff.

With new legislative direction, we went back to the public and held hundreds of meetings and hearings to develop a plan based on grass-roots consensus. In

June 1994 the Coastal Coordination Council published the first complete draft of the Texas Coastal Management Program, and Texas was once again on its way to being a part of the Federal Coastal Program. This Texas Coastal Management Program created no new bureaucracy. It only sought to coordinate existing law.

In Texas, the Coastal Management Program emphasized these policy concerns: protection of wetlands, erosion response and dune protection, access to beaches and other shoreline, oil spill response and other energy resources management, water quality, and dredging. The plan required state agencies to coordinate and streamline their policies and regulations.

In 1990 Rick Perry was elected agriculture commissioner. I'd known Rick since we were students together and he was John Sharp's roommate in College Station. What's more, his office is one floor up from mine in the Stephen F. Austin Building. But Rick Perry's idea of a good coastal program is *Gilligan's Island*. By the time Rick became interested in the Coastal Management Program, there had been about 250 public hearings, coordination council sessions, workshops, meetings, and briefings. All of this discussion and give-and-take was rooted in the fundamental, almost mystical belief of Texans that the beaches belong to everyone. So it's everybody's responsibility to maintain the integrity of that natural heritage. The quest for coherent coastal planning had been about as harmonious as government in Texas ever gets. Then one day in 1994 a *Houston Post* reporter called me and said Rick Perry had just held this news conference blasting me and the Coastal Management Program as big government that was trying to take away people's property rights.

Remember, now, this is a law that was urged upon Texas and other coastal states by Richard Nixon and John Tower. What had changed?

Newt Gingrich called it the Revolution.

15

"Bush's First Big Mistake"

ere I need to go back a little. Twenty-five years, in fact. In 1972, to the dismay of my dad I had temporarily dropped out of law school at the University of Texas and passed up a chance to work in the attorney general's office so I could make $500 a month carrying the bags and driving the car of Ralph Yarborough. Then my candidate blew a million-vote lead in the runoff and I was jobless in June when I had counted on staying busy and paid until November. Now I was really in limbo.

I enrolled in summer school to catch up on my law studies, and got hired by a non-profit group registering new voters. Texas had just passed a constitutional amendment giving eighteen-year-olds the vote. In mid-summer I got a call from a young law student at

Yale working for the Democratic National Committee; she said she had been sent to Austin to register voters. She said she knew our approach was nonpartisan, but their group had very limited resources. She suggested we meet to try to work out an arrangement so we weren't duplicating our efforts. Her name was Hillary Rodham.

I met with her several times, and we got to be pretty good friends. Late that summer she told me that a friend and classmate at Yale was going to be running the Texas Democratic presidential campaign. She wanted to recommend me to head up Young Texans for McGovern, if that was all right. In August Bill Clinton rolled into town, and we set up the headquarters in the same office we'd used earlier for Yarborough. Soon we were friends, too. We weren't kidding ourselves about the odds against carrying the state, but we worked hard, we learned a lot about politics in Texas, and we had a lot of fun.

I always knew my new friends were smarter than the rest of us, but it really struck me one morning in September. Bill and Hillary came running out the door at the headquarters and asked me to give them a ride to the airport.

"Sure," I said. "Where are you going?"

New Haven, they replied. They had to register for classes.

"But...didn't classes start a month ago?"

Right. This was the last day for late registration.

"Well, are you going to stay up in New Haven and go to classes?"

No. They were coming right back to Austin on Monday.

At the time I thought I was a real hotshot. I had a job and had signed up for twelve hours at the UT law school and went to class a limited number of times. I kept up, even if my ranking was in the middle of my class. They were going to Yale and didn't even bother to register on time, much less go to class. And they still graduated in the top ten of their class. And I mean the top *ten people*, not ten percent. That's when I began to understand these folks were in a little different category.

When you get involved in public life, you do it because you want to make a difference. You want to have an impact. And as you serve in office, you develop political capital. You expend that capital to affect things you consider important. And what could be more important than using that capital to help elect a president of the United States?

If I could help Bill Clinton get elected president, think how much more I could accomplish as Texas Land Commissioner. I understood that the Bush family would never forgive me for directing the Clinton campaign in Texas, but on every level, I believed it was the right thing to do.

And two years later, I wasn't going to put distance between myself and my friends just because the First Lady was being vilified and the president had a twenty-five percent approval rating in my home state. But in 1994 I also had to decide what I was going to do. As I said before, I have loved my job as land commissioner since the first day I stepped off the elevator in the Stephen F. Austin Building, and there was still much I hoped to accomplish. But political life has a steep downside, and my close association with the Clintons was just one of my problems.

As land commissioner I had been investigated by the FBI on the basis of an allegation that I was involved in an extremely far-fetched land scheme associated with the veterans loan program. Ultimately the FBI cleared me, the agency, and everyone else impugned in this fiction of all wrongdoing, but at the outset of the investigation, the agent in charge had spoken publicly and rather noisily in my view, using words like "racketeering." During the 1992 presidential race some calls had gone from the Land Office to the Clinton-Gore campaign. Eighty-five percent of them were local calls, and ten percent of them were long-distance calls charged to credit cards, which didn't cost the state anything. But five percent of the calls were a problem. I apologized to the public and to Travis County District Attorney Ronnie Earle and reimbursed the state for the long-distance charges. Afterward, I hired an attorney from the prosecutor's office as an ethics advisor so that nothing like that would ever happen in our agency again. During my third term, I filed personal bankruptcy. And I was a favorite dartboard of the state's largest newspaper, the *Houston Chronicle.*

The Republicans smelled blood in the water, and I had to agree it looked a little pink. I didn't want to run a race for land commissioner I couldn't win. If that was the certainty, I owed it to my staff and all that we'd accomplished to move on and work for a Democratic successor who would respect and build on that record. Also, as land commissioner I was making $79,000 a year. Whatever deadweight it might be in Texas politics, there are distinct advantages to having a law degree, a background in government, and a well-known friendship with the president of the United States. I had several offers from law firms that could start to resurrect my financial standing.

So as crunch time approached for decisions about 1994, my old friend Roy Spence and I retained an accomplished veteran pollster, Jeff Smith, who doesn't do much political work anymore. We wanted to find out what my chances were. And it wasn't one of those polls so common in politics: Tell the client what the client wants to hear. We worded the questionnaire to put me in the worst possible light. We mentioned the FBI investigation, the phone calls, the bankruptcy. We also framed environmental protection policy initiatives as negatively as we could. Assume now that this is going to raise your taxes and cost people jobs...

The results were astounding. On support for strong environmental protection, the lowest we could push the level of support was to fifty-eight percent. And as for Mauro, the respondents said, as long as he's doing his job as land commissioner and protecting the beaches, we'll vote for him anyway!

My opponent in 1994 was Marta Greytok, whom Bill Clements had once appointed to the Public Utility

Commission. But it was her first campaign for statewide elected office, she didn't know much about the Land Office, and since Rick Perry and Railroad Commissioner Barry Williamson figured their own races were in the bag, the Republicans decided the way to get rid of me was with a tag team of naysayers.

The Coastal Management Program got sucked into the campaign because the Republican Party's consultants and bureaucrats in Washington decided that all environmental regulation was vulnerable and that a wedge issue for them was protection of private property rights. The heart of their strategy was the Endangered Species Act.

Through the leadership of his Interior Secretary James Watt, President Reagan had also tried to dismantle this country's environmental laws, but Congress and the people wouldn't stand for it. Eventually the Republicans had the good sense to back off. During the years of Watt's extremism a popular countermeasure had been to give great autonomy to local biologists and other scientists of the EPA and U.S. Fish & Wildlife. Democrats and environmentalists made it a virtue for those bureaucrats to defy their superiors. This became a tradition and, for a time, may have had benefits. But ultimately, twenty-five-year-old bureaucrats with an advanced degree but no common sense were making decisions that affected thousands of individuals, their families, and their livelihoods.

Most people don't understand that the Endangered Species Act primarily regulates public property. You can't kill a whooping crane that lands in your back yard, of course. But the intent of the law was not to regulate private property. Still, I can show you letter after letter from federal wildlife biologists informing a private property owner that something is prohibited because of the Endangered Species Act. Republicans possess a lot of that correspondence, too. Some see endangered species as merely a tool to be used to stop development. They can't get their way in Congress or state legislatures, so they use the Endangered Species Act to get what they want in federal court. The result has been a huge, often irrational backlash against a law that has helped return species like the brown pelican and the bald eagle from the brink of extinction.

During the 1994 campaign, Rick Perry and Barry Williamson used this to great effect on behalf of their gubernatorial candidate, George W. Bush. U. S. Fish & Wildlife communications implied that because of endangerment of two warblers, the golden-cheeked warbler and black-capped vireo, the agency could dictate to rural Central Texans things as personal as where they could build a fence or plant a garden. The organization Take Back Texas became a mighty political force.

But the Republicans, whose spending against me was topped only by what they poured into the governor's race, missed an important distinction in the minds of the Texas electorate. Texans didn't want any form of government telling them what to do with their places in the country—but they wanted their beaches protected!

In their assault on the Coastal Management Program, Perry and Williamson were running all over the state telling voters that I was the candidate of *beach protection.*

My biggest fear was that they wouldn't spend enough money to overcome my long incumbency, my friendship with the Clintons, and all my other negatives. Because their TV attack ads were doing a bang-up job of getting my message across.

On election night, I was getting dressed to go downtown as my children watched the returns on TV. Neal Spelce, a longtime Austin pundit, turned to his co-anchor, Stephanie Williams, and they agreed that I was bound for defeat. David and Alex giggled because they didn't think anybody could beat Daddy. But it was a wipeout for Democrats everywhere in the country. Ann's organization knew from the exit polls that she was going to lose. Longtime Railroad Commissioner Jim Nugent lost. In the comptroller's race, John Sharp got fifty-five percent against an unknown opponent who spent no money. I won my fourth term as land commissioner with fifty-one percent of the vote, and what saved my bacon was my South Texas base and the coastal counties. Without Rick Perry, Barry Williamson, and the Republican attack on Coastal Management, I don't see any way I could have won that race.

But you have to give them credit for being consistent. In December, on her way out of office, Ann became the third Texas governor to sign a Coastal Management Program and the first to send it to Washington for approval. Three months later, Governor George W. Bush sent word to Washington that he wanted the Coastal Management Program sent back.

As a student of Texas politics, I sure respect George W. Bush as a campaigner. He took on a race that few Texans and hardly anyone outside this state thought he could win. He defeated a governor who still had a fifty-seven percent approval rating the day he beat her by 300,000 votes. I admire some of the positions Governor Bush has taken. For example, he pointedly declined to get on his party's bandwagon of blaming every problem in the southwestern states on Mexican immigrants. I remember reading a press account in the *Austin American-Statesman* of how he has worked harder than any recent governor at building a positive relationship with the legislature. I applaud him for that. But in that same story, he was quoted as telling a crowd of legislators that the thing that surprised him most was that they worked such long hours. What a remarkable thing to say!

I've been around a lot of legislators whose positions I strongly disagree with. But it's hard to miss the fact that they're deeply committed to their beliefs and are working day and night.

I'm not sure Governor Bush understood the history and the reasons for the Coastal Management Program. But as a politician he does have constituents, and to those constituents he owes certain debts. He owes the National Rifle Association for delivering votes in return for his backing the legalization of concealed handguns. He owes the Christian Right, whose adherents set the course of education and social services policy; that debt has been honored in several of his appointments. And in March 1995, the ideological attack on environmental protection was at its peak in Newt Gingrich's House of Representatives. Governor Bush owed the ideologues in Texas who used property rights as a camouflage for that agenda. So he listened to his advisers and pulled back the Coastal Management Program. And paid for it.

"Bush's First Big Mistake," editorialized the *Corpus Christi Caller Times.*

"Foolish Action," agreed the *Houston Post.*

Beaumont Enterprise: "Coastal Plan Falls to Partisan Politics."

Under the headline, "Gov. Bush Might Want to Think Again on This One," editorial writers at the *Houston Chronicle*—hardly cheerleaders of mine in the past—scolded the governor: "To scuttle the plan now is not a good idea, especially if the goal is indeed local control. Land Commissioner Garry Mauro is correct in pointing out that federal approval of the plan would 'allow Texans to run Texas.'"

The overwhelming support for the program led Governor Bush to promise to resubmit the plan after the legislature made a few changes to it. In the spring of 1995, a bipartisan coastal bill—cosponsored by Democratic Senator Carlos Truan of Corpus Christi, Republican Senator Jerry Patterson of Pasadena, Democratic Representative Hugo Berlanga of Corpus Christi, and Republican Representative Steve Holzheauser of Victoria—overwhelmingly passed the legislature. The big change the legislature made was to add three agencies and two more coastal residents to the Coastal Coordination Council. Other than that, they changed very little of what coastal residents had put together since that first advisory committee meeting in 1990.

The controversy went on all through the summer, however. In June, Railroad Commissioner and Coastal Coordination Council member Barry Williamson proposed rules that made sweeping changes to the program. This prompted Jeffrey Benoit, the director of the federal Office of Ocean and Coastal Resource Management, to write the governor a letter. It said in part: "The proposed rules replace relatively straightforward state consistency review procedures with needlessly cumbersome, bureaucratic and arbitrary procedures resulting in a process that is ineffective and biased toward maintaining the status quo."

Following an August public hearing in Galveston in which frustrated coastal residents blistered the Williamson proposals, Governor Bush publicly agreed that the rules had to be changed to make the program acceptable to the federal government. We made some changes that Governor Bush, Commissioner Williamson, and members of the legislature wanted. And on January 10, 1997, twenty-five years after John Tower's legislation advanced the idea, Texas coastal residents got the planning mechanism they had been denied so long.

In March 1997 we held a conference in Corpus Christi to bring coastal residents together to continue the planning and just to celebrate. We recognized organizations, industries, and individuals—especially two pioneering champions of the Texas coast. Presented by the dapper old congressman himself, the first Bob Eckhardt Lifetime Achievement Award went to Babe Schwartz.

We're told that our colleagues in Washington are amazed at how we've utilized the grant money. In most states, the funds are used to pay the salaries of bureaucrats. Here we've passed on ninety percent to the local communities—something no other coastal states do now, but they're starting to rethink that. It's not a lot of money, but it has been used for things like the Pirates Beach dune reconstruction in Galveston; making a public boat dock in Port Arthur a recreational feature of urban revitalization; and helping Boy Scouts build a new pavilion on Padre Island.

It's not a perfect Coastal Management Program. Nobody is entirely satisfied with it. But we protected the plan's essential elements—maintaining beach access, erosion control, and wetlands protection. We also helped to achieve a local consensus. The Republicans engineered a powerful majority on the Coastal Coordination Council. But you know how it's turned out? Even though there was a concerted effort to undermine the program by political consultants, in the end the Republican citizen appointees have voted just like their Democratic predecessors—to protect the Texas coast for all time.

16

For All Texans, For All Time

In any calling, what you do is not all that's important. What you do is not even the most important. It's what you believe in. How you treat people along the way. And what you learn from them.

When I first decided to ask the people of Texas to elect me land commissioner, I knew I had some powerful learning to do. Everyone who lives in Texas eventually falls in love with the coast. And in ways I didn't fully understand, the coastal heritage lay at the heart of the office I had decided to seek.

In 1980 I went to see one of the great figures—and great personalities—in twentieth century Texas, Babe Schwartz, to tell him that I'd decided to run for land commissioner. As the state senator from Galveston, Babe carried on a valiant and unrelenting fight to protect the coastal heritage from all the forces that could so easily and, in many cases, unwittingly destroy it. In most other states, the rich and powerful had been able to privatize the beaches and fence them off into enclaves of privilege. Against all odds, almost in spite of ourselves, that never happened in Texas. Thanks to defenders like Babe Schwartz, the beaches, up to the vegetation line, belong to everyone.

And Babe's passion for coastal protection wasn't limited to his legislative bills and orations. It has been a large and consuming part of his life. Babe must have spoken to all the coastal citizens' associations. He must have read and dissected every government white paper. To me it was almost like he knew every dune and beach. I wanted his perspective.

As I sat in his office and we talked, my gaze kept falling on a piece of art he had on his wall. The framed picture bore a striking likeness of that towering nineteenth-century Texan, Sam Houston. It was adapted from a famous painting in which the hero of the Battle of San Jacinto and the first president of our republic struck a somber and rather intimidating pose. Below that was a inscription. "I have but one maxim: Do right and risk consequences. Sam Houston."

"Babe," I finally said. "That's a great picture."

"Good," said Babe, as if I'd absorbed the most crucial thing he could tell me. Babe had just lost his seat in the Texas Senate, and Ronald Reagan had just been elected president. Babe rose from his chair, took the piece of art off the wall, and inscribed it: "With great hope for the motto to prevail in an otherwise bleak time in our nation." Then he gave it to me. In January 1983, when I walked into the Land Office—my first day on the job—that framed picture was all I carried with me. It's been on the wall behind my desk for almost fifteen years.

I've not always honored the advice of Sam Houston or, for that matter, Babe Schwartz. And when I haven't, that's when I get in trouble. Those words of Sam Houston are the needle on my life's compass. When I stray off course, it never turns out to my advantage. I wind up disappointing myself and everyone around me. But when I risk the consequences and do what I know is right, I've found I can accomplish a great deal that's good.

My office is in the building named for Stephen F. Austin, the father of Texas who loved this land enough to risk the consequences of leaving the United States and committing his life to being a citizen of Mexico on its most unsettled and dangerous frontier. My office has a great view south toward our pink granite capitol, that monumental building whose construction was financed by the transfer to cattle barons of three million acres of public land.

The panoramic view of the Capitol and the green Hill Country beyond has been one of the simple and sustaining pleasures of my time as land commissioner. I look across Congress Avenue at another government office building named for Lyndon B. Johnson. Because of Vietnam, the lessons learned from him are complicated, to say the least—especially for my generation and gender. But here was a public school teacher who emerged from a small-town upbringing in the Hill Country and became the American president whose enduring legacy is civil rights. "We shall overcome," he drawled the song lyric that had become the rousing call to action of Martin Luther King., Jr. That was LBJ's finest hour, and our nation is better and stronger for it today.

When I think of Lyndon Johnson it brings back memories of his colleague and sometime adversary, Ralph Yarborough. Senator Yarborough was a white-knuckler on an airplane; if there was any way around it, he refused to fly. All those years ago, when I was driving Senator Yarborough all over Texas in his last campaign, he told me things at two in the morning he never would have said in his office. In the Senate he had given future generations of Texans the Padre Island National Seashore, the Big Thicket National Preserve, the Guadalupe Mountains National Park. But all his political life he had wanted to be governor of Texas. Until the moment a decision could no longer be delayed, he considered making the governor's race in 1972. I think that's what he really wanted to do. But because of Vietnam and the ominous temper of the Nixon presidency, old colleagues in Washington convinced him that the country needed him back in the U.S. Senate. So he gave it one more try and failed. He took the loss like a champion.

Ralph Yarborough was a hard man to befriend because he held himself and everyone else to such high standards. When his friends failed to measure up to those standards, he often judged them harshly. After Senator Yarborough died, his wife Opal gave me a photograph that President Johnson inscribed to his friend Ralph Yarborough. I have that on a wall in my home. Late at night twenty-five years ago, Senator Yarborough used to talk about the difficulties but also the private warmth of that relationship, which was so often viewed by the public and press for its elements of conflict. The pragmatist Lyndon Johnson was always trying to compromise and attain the art of the possible; the idealist Ralph Yarborough always tried to be uncompromising and create an art of overcoming the impossible. That made for a tense and prickly friendship, but together they accomplished a tremendous amount for the people of Texas. As an elected official, I've always wished I could combine the qualities of both men.

The only way you can accomplish things in public life, in my opinion, is to understand how we got where we are. If you don't have that understanding, you can't make improvements, and you can't avoid the mistakes made before you. The land commissioner's job is to run a trust fund for education that has been in existence a hundred years. If you don't understand how that trust fund got where it is today, chances are slim you can protect it so that it will be here in another hundred years. You've got to have a sense of history. And just outside my window is so much history, so much Texas.

I think of another mentor, Bob Bullock, and the

years I worked for him in the comptroller's office. Long ago, Mr. Bullock—as we all called him, carefully—thought about running for land commissioner. He was the one who taught me what the Land Office does: where it fits in the complex machinery of Texas state government, and how important it is in ensuring that our state has a sound and well-financed system of education.

Bullock changed state government in Texas fundamentally for the better. He demonstrated that it can be a modern businesslike operation, run without waste, but also responsive to the needs of all Texans. He proved it all those years directing the comptroller's office, and you notice that when he got elected lieutenant governor, from the first day of his first session in the senate chamber, he showed he had the legislative talent of a Sam Rayburn or Lyndon Johnson. That's a very different arena, and a hard transition to make. Bullock believes that public service is a high calling, and he's a hard taskmaster. He'd tell us that there are two kinds of government employees. One who says, Well, the law doesn't allow us to do that, or there's a chance that won't work, or we've never done that before. Then there's the kind who reads the same law and realizes what *can* be done to help the people who pay our salaries.

I learned something else important from the philosophy and example of Bob Bullock. What matters most is substance. Yes, appearances are important. But substance is what lasts. What you've got to think about is what's right for tomorrow. What's right for ordinary people who don't think much about politics and government. They're just out there trying to provide for their families and make a living, have a decent quality of life. Do right by them, and the politics will take care of itself.

Do right and risk the consequences.

One of my colleagues at the comptroller's office was my old friend and fellow Aggie, Tom Henderson. When I met Tom in College Station he was a pre-veterinary student and prominent member of the Baptist Student Union. You can't believe how competitive the pre-vet program is at Texas A&M; his grades were at the top of his class. Later, when I was running for president of the student body, Tom was supposed to be my campaign manager. But he was also on the campus election commission, and when he found out that my opponent didn't have the grades to qualify, and that the administration wasn't inclined to get involved in the matter, he was outraged. This was against the rules! My campaign was in shambles, and all anybody was talking about was whether Al Reinert should be disqualified. Wasn't exactly what I had in mind. I wanted to talk about what we needed to do at Texas A&M.

In 1971 we went over to Waller County with Ed Wendler—an Austin lawyer who became a lifelong friend—as activists of the Texas Student Legal Defense Fund. The state had passed a law allowing college students to vote in the counties where they attended school; before then, they could only register in their home counties. But Waller County officials refused to honor the law. They knew that black students enrolled at Prairie View A&M could disrupt the status quo if they voted. Students were thrown in jail for trying to register. Through the bars, they were given shavings of soap to wash their faces in the morning. That voting rights campaign was how Tom and I got to be friends with Mickey Leland, a lifelong champion of voter's rights, and Craig Washington, who was the students' attorney. I'll never forget that day on the courthouse steps when the sheriff met us and the students trying to register with his hand on the grip of a holstered, very large gun.

In his professional life, Tom's a sprinter. He's absolutely driven by what he believes in, and he'll work harder and be more single-minded about achieving a goal than anyone I've ever met. But Tom's taken some hard falls in his life. After we left the comptroller's office, he realized he's gay, and he had to admit it. His successful law practice evaporated almost overnight. One day in 1986 he called me and said: "Hate to do this to you, old buddy. But I need a job. Bad."

In 1987, along with all the bulldog persistence Tom Henderson provided on the quest for ratifying the MARPOL treaty and his vision of linking clean air and Texas natural gas, he helped me design and implement the first written policy in any Texas state agency prohibiting employment discrimination on the basis of sexual orientation and HIV/AIDS. And we set up mandatory training on AIDS awareness. That's particularly important to Tom because he's HIV-positive, and it's the right thing to do.

Bill Clinton, our old friend from that losing presidential campaign twenty-five years ago, made Tom a member of the Presidential Advisory Council on HIV/AIDS. That doesn't have one thing to do with Tom's job at the Texas General Land Office, but it's important work.

You do what's right. You don't turn away from your friends.

I learned a lot from my grandfather, who had a third-grade education.

After the 1982 Democratic primary, when I was running for my first term, it was two-thirty in the morning, and my campaign headquarters, set up in my law office, was like a morgue. All my friends were calling to console me; every newspaper in Texas had me running third and missing the runoff by 25,000 votes. I elected not to believe the *Houston Chronicle* and *Dallas Morning News* because so many votes in South Texas were still out. I was the first statewide candidate to campaign on the premise that if you carry South Texas overwhelmingly, it's hard to lose a Democratic primary.

On the other hand, at four in the morning it's hard to stuff a pillow on the possibility that you might be in denial. I didn't get half an hour's sleep. At ten I had scheduled a meeting with members of my finance committee; they included my old friends Roy Spence, Ed Wendler, Leon Thompson, Billy Goldberg, and my cousin Don Mauro. They all showed up to offer their condolences. Instead I was asking them how much money they could commit to this runoff I was certain that I was in. Basically what I did was ask them to sign notes at banks. They committed to $400,000, and I'm convinced that at least half of them did it just to get me through this difficult hour and ease my youthful pain.

It turned out I was right about South Texas, and the race against Pete Snelson, the state senator from Midland, was on. In the primary I had run a weak third in East Texas. Roy Spence was doing the advertising, and we were putting together a mail piece that was supposed to help me do better in the eastern part of the state. We were doing this at my Grandfather Mauro's house in Bryan. I had in mind a back cover that emphasized my connections to Texas A&M.

My grandfather said: "Well, you know I'm from Bryan, and A&M is a real big deal here in Bryan, but seems like if you didn't go to A&M, you might care about some other school. Every Aggie you meet, there's ten who went to Baylor or Stephen F. Austin or some other place. I'm not sure I'd do that."

Roy agreed. "I'd rather do a piece on your family. Let's get some pictures of your parents, your great-grandparents—tell the story about how you came from Sicily. You know, the log cabin story. Everybody can relate to that." Roy hesitated. "Besides, Mauro is a strange name. What kind of name is *Mauro* anyway?"

The quiet in the room was uncomfortable. Without spelling it out, we were talking about what assets or prejudice a name implies.

Mauro is a common given name among people of

Hispanic descent; it's less common as a surname. In one border county, a local official with that first name for years discreetly borrowed my campaign signs. But in that first race I noticed that when I made a campaign stop, people would often ask what church I belong to. As it happens I'm Catholic, and that used to be a political liability. But these potential voters were trying to pin down my ethnicity.

Roy and I were talking delicately and I suppose a little cynically about how best to leverage my "different kind" of name in different parts of the state.

My grandfather told me once that he quit school in the third grade because children of Italian and Hispanic families were segregated from the white students into a separate classroom. The teachers spoke English, the Sicilian kids spoke Italian, the Hispanic kids spoke Spanish; and nobody could understand each other. And they divided off in groups in the playground. Public education wasn't getting him anywhere; he just gave up.

My grandfather listened to us plan our political strategy and at last he had to speak out. "You boys are college boys, and you all probably know a lot more than I know. But Garry, I don't think those kind of people like Italians any better than they like Mexicans. I wouldn't put that in your mailing."

I learned then from my grandfather that liabilities come attached to all things. You accept them and go on. But at the same time you had better know who you are, and be proud of who you are.

In politics I've chosen to be a Democrat. Because in

my mind, Democrats employ government as an equalizer. Government is supposed to make certain that every person in America has a chance to reach his or her full potential. Call it a cliché of the Clinton presidency if you want: But if you work hard and play by the rules, you ought to have a chance to fulfill your dreams and know that your kids will have a decent education, and if they work hard and play by the rules, they'll have a better standard of living than their parents and their grandparents did.

In my mind, Republicans are almost always protecting the status quo or looking toward the past. That's not to say Democrats haven't gotten way out of line and out of touch, on occasion. For one thing, Democrats have never figured out how to control the bureaucrats. That's nothing new, though. Corporations can't control their bureaucrats either. I am a devout believer in free enterprise. But I can't believe the private sector could have given us voting rights and civil rights or desegregated the schools. Or that corporate America would ever have made it a priority to protect the environment and enforce workplace safety standards. Or provided affordable housing, a G.I. Bill, and health care for military veterans.

I know where I came from. How could I not be a Democrat? We were tenant farmers. If it hadn't been for the New Deal in the thirties, my great-grandfather would never have been able to buy his 180 acres. If it hadn't been for the G.I. Bill, my dad never could have gone to college. If it hadn't been for student loans, I never would have finished undergraduate and law school. If it hadn't been for the principles of the Democratic party, someone of my social and ethnic background could never have been elected to a constitutional office in Texas. I know how I got where I am. It's easy to remember

Do right and risk the consequences.

When I first got involved in politics and government, the standard advice and common wisdom of political consultants—in Texas anyway—was that environmental issues are a loser. Environmental issues are dangerous. Environmental issues are politically unpopular, particularly with people who finance campaigns. And, to tell the truth, if I had to pick ten words to characterize myself today, "environmentalist" would not be one of them. I started out thinking environmental concern was largely an issue of aesthetics. But in the last fifteen years, I've learned from ordinary Texans that they want clean air, clean water, clean beaches. I've learned from Texans that they don't want tradeoffs. They want a thriving economy, a quality of life that enriches the upbringing of their children, and a clean environment. Because the bottom line is human health. Family health. The health of their old ones. The health of their infants.

And so I've taken stands and carried on fights that aren't going to get me appointed to many boards of corporations. Education. Economic security. Equal opportunity. And the environment. Those have been the guideways of my time in public office, though I doubt I could have articulated that, when I was a young man just starting out. Those are things worth fighting for, and accepting whatever risks come along the way.

I'm not a poet and I don't write essays. I'm a forty-nine-year-old father of two of the greatest kids on earth. I'm someone who for fifteen years has loved to get up and go to work. I don't know how much else you can ask for. I'm proudest that during my time as Texas Land Commissioner, we've managed to rewrite the laws that deal with our public lands and natural resources in a way that probably protects them for future generations better than those of any other state. It's mostly arcane stuff, but that doesn't dim my satisfaction. This evening, though, I'm not working. I'm in Port Aransas, Texas—what a marvelous place to live! Right now, it's snowing in Amarillo, but it's eighty degrees on Mustang Island. Gulls are hanging, almost motionless, in the rich salt breeze. I'll bet I could find a sand dollar. The beaches are cleaner, and I'm going for a walk.